Race Car Braking Systems

By Steve Smith

ISBN No. 0-936834-07-2

Steve Smith Autosports

P.O. Box 11631, Santa Ana, CA 92711

Revised August, 1980

NOTICE

This book is expressly written to assist in the selection and use of race car braking systems and components, and to familiarize the reader with all the basic concepts of the braking system. All of the part numbers and recommendations given in this book are presented without any guarantees on the part of Steve Smith Autosports and the authors. The authors and publisher disclaim any liability incurred in connection with the use of this data or any specific details contained in this book. Because the contents of this book deals with matters highly critical to the safe operation of a racing vehicle, we strongly urge the reader to study each section of the book carefully before applying any recommendations or specific parts to a race car.

A SPECIAL THANKS

In digging out all of the information contained in this book, we encountered a very great group of dedicated and helpful people. We would like to say a special thank you to Marvin Panch—Grey Rock, Lee Hossler—Grey Rock, Dale Dildine—Grey Rock, Mac Tilton—Tilton Engineering, Paul Lamar, Don Alexander, Frank Deiny—Speedway Engineering, Tim Manion—Bendix, Bill Howell —Chevrolet, and Larry Rathgeb—Chrysler.

TABLE OF CONTENTS

How the Braking System Works

How the master cylinder feeds pressure through the braking system is a basic lesson in hydraulics.

The system begins with an actuating force fed into the master cylinder. This force is the driver's foot on the brake pedal. The most force that a driver can generate is between 150 and 200 pounds, with most racing brake applications being about 100 pounds of force. The driver is also limited to about six inches in the amount of pedal travel he can apply (depending on the driver's leg length, closeness to the pedal and leg angle). The ability of the driver to apply force to the pedal is determined by body restraints (seat belt and shoulder harness), and the angle of the hip, knee and ankle joints. The more these three joints are in a staight line, the easier it is to create high force on the pedal. The force from the foot is then multiplied by the leverage of the pedal lever on the master cylinder push rod. On most passenger car applications this leverage is between 3 to 1 and 4 to 1. This leverage could be as high as 7 to 1 with a floor mounted master cylinder. If 100 pounds of force is applied to the brake pedal and the leverage ratio is 3 to 1, then the input force into the master cylinder is 300 pounds. Remember that if the pedal ratio is increased, the force will increase, but the pedal travel will also increase. A good idea would be to have alternate holes drilled in the pedal lever so alternate leverages can be applied—much cheaper than changing master cylinders.

The master cylinder bore size (and thus its square inch area) is the next consideration. The most common master cylinder bore size used in racing vehicles in one inch. This gives an area of .79 square inches (area equals pi times the radius squared). The force supplied into the master cylinder is *divided* by the square inches of the piston, and the output of the master cylinder is pressure in pounds per square inch.

From this knowledge, it is easy to see that the *smaller* the master cylinder bore, the more pressure which is supplied into the brake lines to the wheel cylinders. For example, if 150 pounds of force were applied to a one-inch bore cylinder, the line pressure would be 191 pounds per square inch. If that same 150 pounds of force were applied to a 7/8-inch bore cylinder, the line pressure would be 250 pounds per square inch. So why not use a smaller master cylinder bore, such as a 7/8-inch? There are two reasons against it: 1) It requires more pedal travel for a given fluid displacement, 2) Higher line pressure aggravates line and caliper expansion and thus uses up more fluid. If a system is so stiff (meaning no or little line expansion or caliper deflection) that the

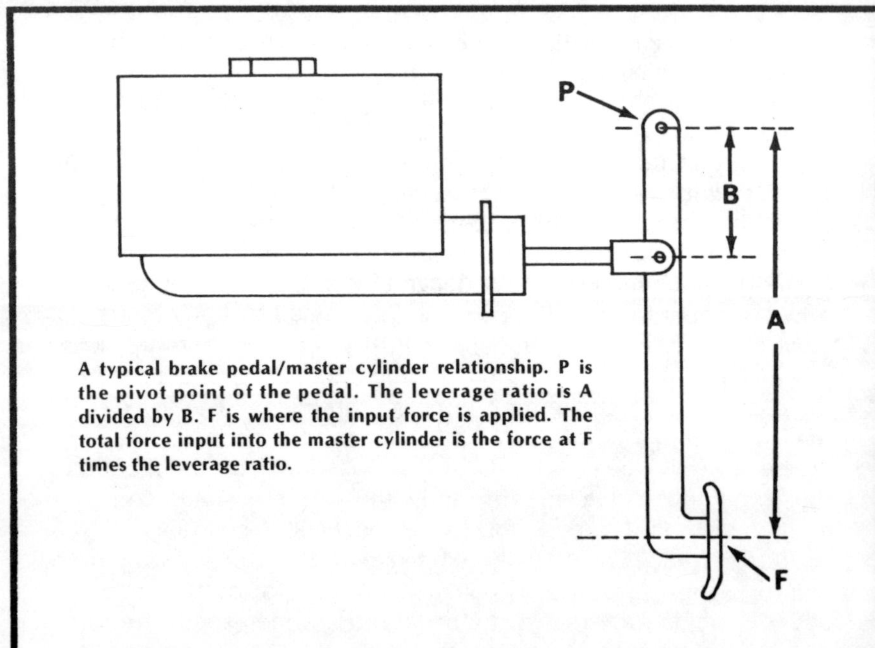

A typical brake pedal/master cylinder relationship. P is the pivot point of the pedal. The leverage ratio is A divided by B. F is where the input force is applied. The total force input into the master cylinder is the force at F times the leverage ratio.

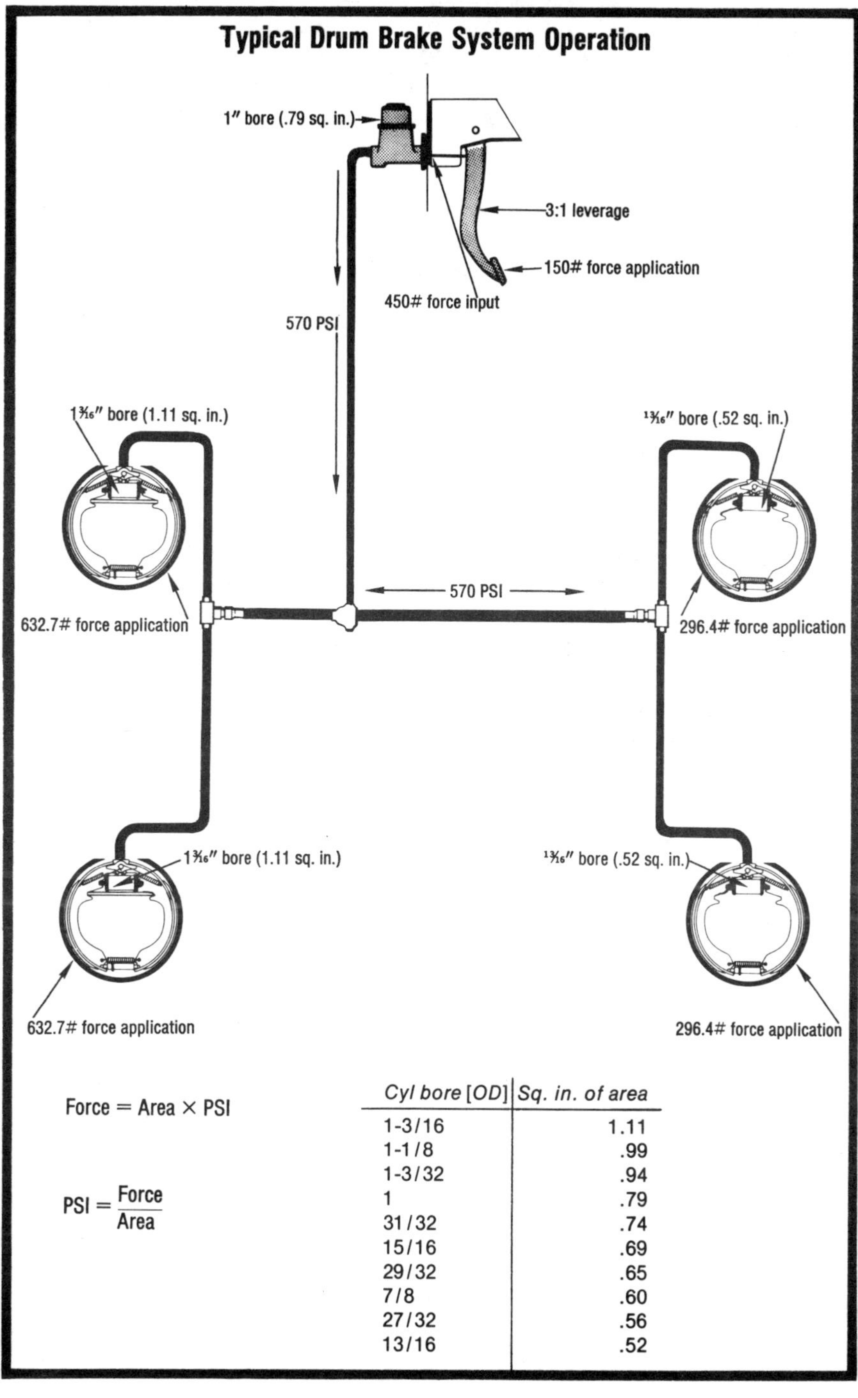

Cyl bore [OD]	*Sq. in. of area*
1-3/16	1.11
1-1/8	.99
1-3/32	.94
1	.79
31/32	.74
15/16	.69
29/32	.65
7/8	.60
27/32	.56
13/16	.52

driver is only using a small part of the total pedal travel, then a master cylinder with a smaller bore will reduce the pedal force, or effort required by the driver, to stop the car.

Once the pressure leaves the master cylinder, there is no great movement of fluid. There is some, but what mostly is happening is pressure transmittal. You can experience this principle by blowing through a drinking straw. Keep your finger a couple of inches from the end of the straw as you blow air and you can feel air movement. Now, block the end of the straw with your finger. There is still a volume of air in the straw, and you are still generating force by blowing into the straw. But your finger only feels force against it. This is the same principle of pressure transmittal as used in the brake system.

THE BRAKE LINES

The brake lines function as the blood system of the braking system, and just as in the human body, there are several pertinent facts which should be remembered to permit good circulation.

Never route brake lines near any source of heat, such as headers or exhaust pipes. Also, never route the brake lines in such a way as to have a loop or kink or high point where air can get trapped and defy bleeding techniques.

Use steel brake lines as much as possible, keeping lengths of flex lines to a minimum. Always use thick wall steel lines, never any other material. Steel has the highest modulus of elasticity of any material you can find for lines, thus helping to prevent line expansion.

You will notice that we have dwelled on the rigidity of the brake lines. This is important if you want the force you impart to the master cylinder to be available at the wheel cylinders. If you use

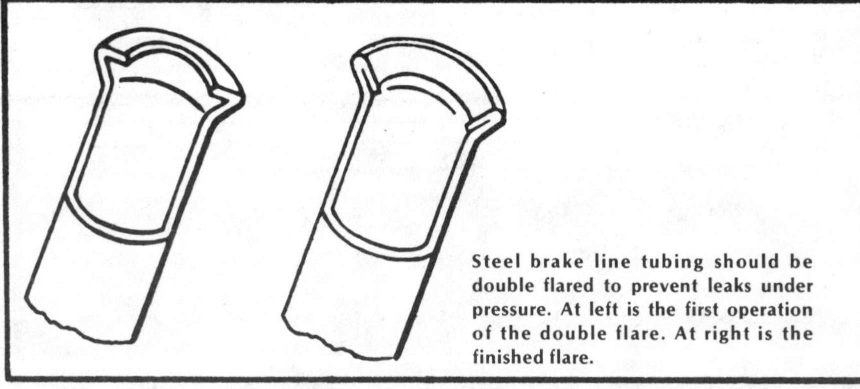

Steel brake line tubing should be double flared to prevent leaks under pressure. At left is the first operation of the double flare. At right is the finished flare.

that force in expanding the brake lines or calipers instead of applying force to the linings, you cannot expect to stop the vehicle effectively, and the driver will merely use all the available pedal travel. The pedal will bottom-out on the floor or the master cylinder piston will bottom in its bore. By making the system as rigid as possible, you are going to get the work to the linings, where it belongs.

Many people believe that they can proportion braking force from front to rear by installing larger diameter lines to the front wheels than to the rear. This is a fallacy. Pressure is independent of volume or area, so a 3/16-inch I.D. line to the front wheels will deliver the same pressure (PSI) as a 1/8-inch I.D. line to the rear wheels.

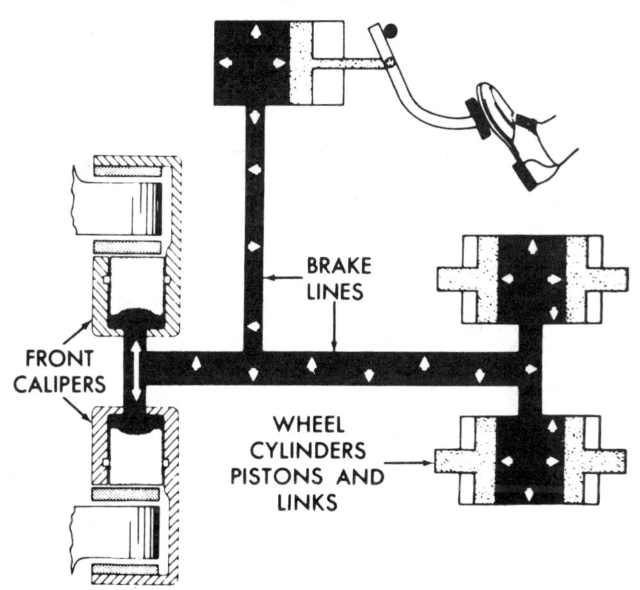

Hydraulic pressure is distributed equally to all wheel cylinders regardless of distance or difference in line size or cylinder size.

The rigidity of the lines we have talked about should be carried into the flex lines. Only stainless steel braided flex hoses can carry the fluid from the steel lines to the wheel cylinders with the positive assurance that little pedal travel or work effort will be lost through line expansion. This is especially important in disc brake systems where line pressures are much higher than in drum brake systems. The flex line to use is Aeroquip 2807-4, which has an

Assembly Instructions for Aeroquip 2807 Hose of Teflon *with* "super gem"® *Reusable Fittings*

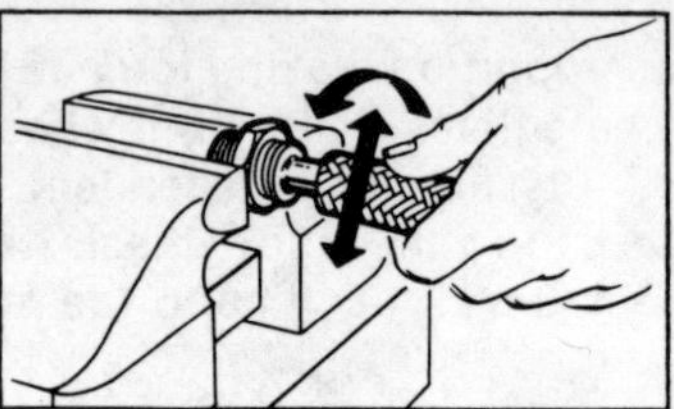

Step. 1. Wrap hose with masking tape at cut-off point and cut squarely to length through taped area using a sharp cut-off wheel or a fine tooth hacksaw. Remove tape and trim any loose wires flush with tube stock. Any burrs on the bore of the tube stock should be removed with a knife. Clean the hose bore. Sometimes wire braid will tend to "neck down" on one end and "flare out" on the opposite end.

This is characteristic of wire braid hose and can be used to an advantage in the assembly of the **"super gem"** Fittings. Slip two sockets back to back over the "necked down" end of the hose, positioned approximately 3 inches from each end. Mount nipple hex in a vise. Work the hose bore over the nipple to size the tube and aid in separating the braid prior to fitting the sleeve. Remove hose from nipple.

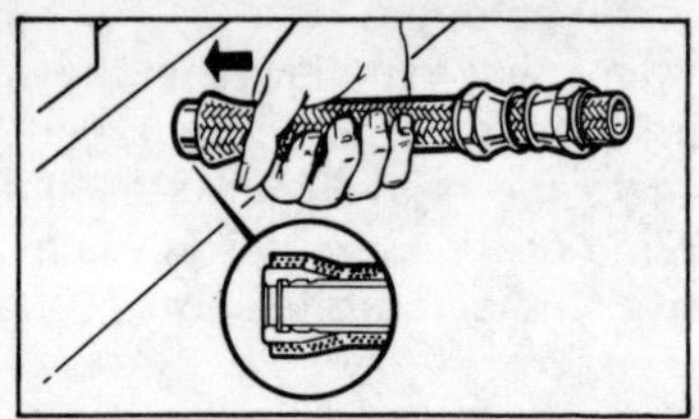

Step 2. Push the sleeve over the end of the inner tube and under the wire braid by hand. Complete positioning of the sleeve by pushing the hose end against a flat surface. Visually inspect to see that tube stock butts against the inside shoulder of the sleeve. Set the sleeve barbs into the Teflon tube by pushing the assembly tool or a round nose tapered punch into the end of the sleeve and tube.

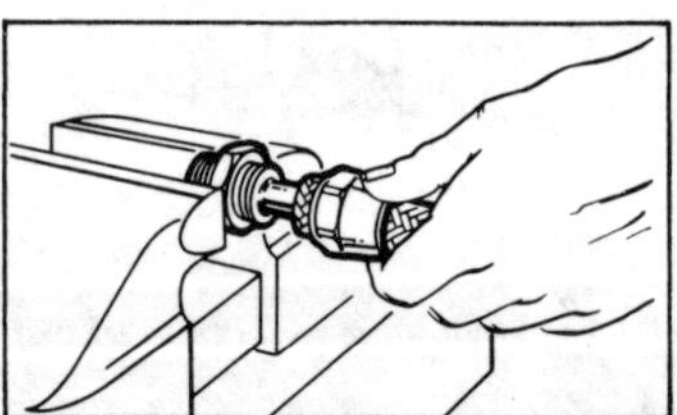

Step 3. Lubricate nipple and socket threads. For stainless steel fittings, use a molydisulfide base lubricant (e.g. Molykote Type G); lubricants containing chloride are not recommended. Other material combinations use standard petroleum lubricants. Hold the nipple with hex in vise. Push hose over nipple with twisting motion until seated against nipple chamfer. Push socket forward, and hand start threading of socket to nipple.

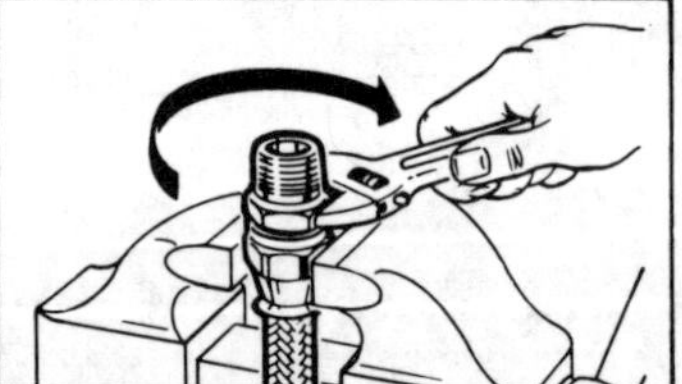

Step 4. Wrench tighten nipple hex until clearance with socket hex is 1/32" or less. Tighten further to align corners of nipple and socket hexes. CLEAN, PROOF TEST TO TWICE OPERATING PRESSURE, AND INSPECT ALL ASSEMBLIES.

to disassemble

Unscrew and remove nipple; slide socket back on hose by tapping against flat surface; remove sleeve with pliers.

I.D. of .19-inch and a working pressure rating of 1500 PSI. It is also quite stiff. The interior hose is extruded teflon, which is a perfect material for brake fluids. Note that many different types and ratings of Aeroquip hose appear the same on the outside, so be sure you are getting 2807-4. This number should be applied at regular intervals on the hose. If you want to purchase this flex hose in pre-trimmed lengths with the correct fittings already attached for your system, they are available from Earl's Supply in Lawndale, Calif. See back section of this book for address and phone number. Aeroquip hose is also available at Caterpiller dealers.

A final fact about brake lines: never assume that any brake lines or hoses you purchase are clean. Pour clean brake fluid through them to be sure the smallest amounts of moisture and dirt are removed.

WHEEL CYLINDERS

The wheel cylinders are very simple pieces of cast metal with a bore in them. In the bore are two rubber cups, each with a piston. As brake fluid under pressure is admitted into the wheel cylinder in the center, it drives the pistons outward equally. The pistons are connected to the brake shoes by a rod link.

Wheel cylinder bores, just like master cylinder bores, come in differing sizes in order to proportion the amount of braking force being applied. Because force (the output desired from the wheel cylinder pistons) equals area (the square inches of the piston) times pressure (the PSI of the brake fluid), the larger the wheel cylinder bore, the greater the force applied to the brake shoes.

Pay attention to the fit of the piston in the wheel cylinder. Too large a discrepancy will allow the cups to partially extrude and thereby use up some extra pedal travel. "O" rings might work better than regular brake cups because brake cups are thick rubber and therefore compress. This can apply to the design of the master cylinder as well (in fact, the Edco master cylinders employ O rings).

For correct brake proportioning on most stock cars, the starting choices for wheel cylinder sizes are 1-1/8 inch bore front cylinders and 13/16-inch bore rear cylinders. Variations on these choices are 1-3/16-inch cylinders in the front with 7/8-inch cylinders in the rear, and 1-3/32-inch cylinders in the front and 13/16-inch cylinders in the rear.

A complete listing with source and part numbers will be found in the accompanying chart for most all sizes of wheel cylinders required.

WHEEL CYLINDER SOURCES

Auto source	*Cyl. bore*	*Part #*	*Manufacturer*
'70-'72 Chevelle	1-1/8	37112-13*	Grey Rock
'68-'72 Nova	1-1/8	37019-20	Grey Rock
'69-'70 Buick	1-3/16	37146-47	Grey Rock
'67-'68 Cadillac	1-3/16	36041-42	Grey Rock
'68-'70 Olds 88 & 98	1-3/16	36041-42	Grey Rock
'65-'70 Chevy	1-3/16	37078-79	Grey Rock
'70-'71 Chrysler	1-3/16	37171-72	Grey Rock
'69 Merc Montego**	1-3/32	37118-19	Grey Rock
'67-'68 Mustang (390 eng.)	13/16	36076	Grey Rock
'65-'70 Chevrolet	1	37080	Grey Rock
'70 Merc Montego wagon	31/32	37161-62	Grey Rock
'69-'70 Buick Riviera	15/16	37114-15	Grey Rock
'70-'71 Chrysler	15/16	37235	Grey Rock
'70 Merc Montego sedan	29/32	36019-20	Grey Rock
'70-'72 Buick Special, GS	7/8	37024	Grey Rock
'70-'72 Chevelle	7/8	37024	Grey Rock
'69-'70 Cougar**	7/8	17507-08	Grey Rock
'72 Merc Montego, 6 cyl.	27/32	37108	Grey Rock
'67-'68 Mustang (390 eng.)	13/16	36076	Grey Rock
'66-'72 Ford Bronco	13/16	36076-77	Grey Rock
	1-1/8	EW 39948-49	EIS
	1-1/16	EW 39950-51	EIS
	15/16	EW 9150-51	EIS
	13/16	EW 39958-59	EIS

* First number indicates left hand part. Second number, such as 37113, indicates right hand part.
**With 351, 390 or 428 engine.

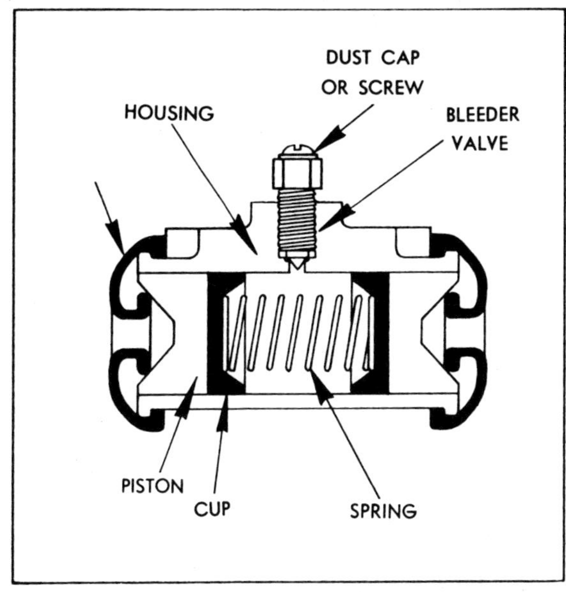

Cutaway drawing of typical wheel cylinder. Drawing courtesy of Grey Rock.

BLEEDING

Once the braking system has been installed, all that is required is to add fluid into the master cylinder reservoir and pump the brake pedal to distribute it into the system. As the fluid travels through the system, it is displacing the air which filled the system. When the fluid reaches the wheel cylinders, there will be a large amount of accumulated air which must be released. This is the reason for the bleeder screw, which is usually fitted at the highest point of the wheel cylinder.

To bleed the system, open the bleeder screw with a box wrench and attach a hose over the fitting and run it into a jar containing new brake fluid. Be sure the tube is well covered with fluid. Start with the farthest wheel from the master cylinder, which is usually the right rear, and progressively work to the closest wheel cylinder. Have an assistant press the brake pedal to the floor. Air and fluid will escape in spurts together. Make sure your assistant returns the brake pedal very slowly. Continue the process until all air has escaped and a solid column of fluid is observed. Then move to the next wheel. Be sure to check the fluid level in the master cylinder after each wheel has been bled. If a bleeder hose and jar are not used as we describe, be sure to close the bleeder screw before your assistant returns the pedal. If not, the system will inhale air through the bleeder screw.

With disc brake systems where air bubbles are more difficult to

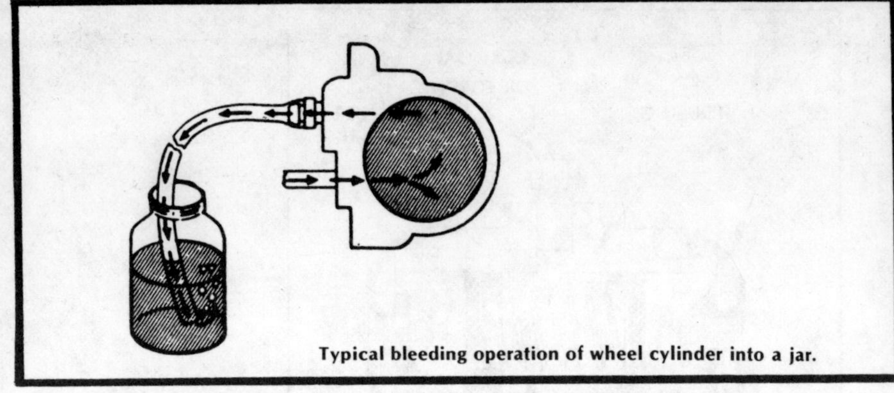
Typical bleeding operation of wheel cylinder into a jar.

dislodge, it is a good idea to use a plastic or wooden mallet to tap the calipers to assist the air bubbles. It is also a good idea to bleed disc brakes once a few practice runs have been made with a new or just-repaired system. With disc brakes, a weekly bleeding is a good idea.

SPONGY PEDAL

If you have faithfully bled the system and the pedal is still spongy, like air is still trapped in it, check out these suggestions: (1) leaks in the system, (2) worn wheel or master cylinder seals, (3) insufficient fluid in the master cylinder, (4) brake fluid is boiling at some point in the system.

THE MASTER CYLINDER

The master cylinder is not much more than a simple hydraulic cylinder where force (measured in pounds) is fed into a piston (measured in square inches) sliding in a cylinder, creating hydraulic pressure (measured in pounds per square inch).

A feature of the master cylinder on drum brake systems is the residual pressure check valve. This valve maintains a small amount of pressure in the brake lines to prevent air from being drawn in through the wheel cylinders and to prevent wheel cylinder pistons from being completely retracted. Disc brake systems will not function properly with this check valve installed in the master cylinder.

Master cylinders and pedals must be securely mounted on a body or chassis structure where pedal force will not move the master cylinder. If it is not rigid, it has the same effect as having the brake lines expand.

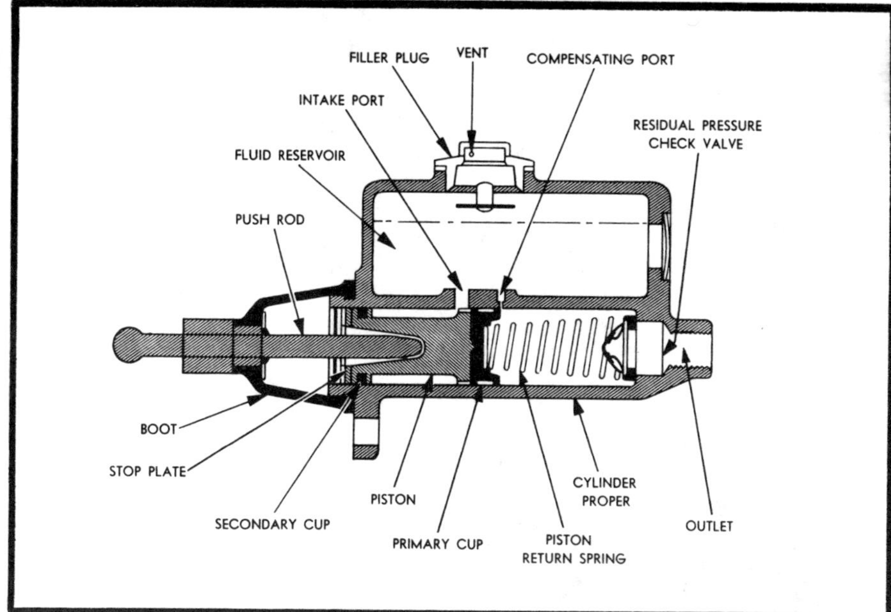

The typical single piston or single reservoir master cylinder. The push rod forces the piston forward in the cylinder bore. The sealing mechanism ahead of the piston is called the primary cup, and as this travels down the cylinder bore it closes the compensating port so that the hydraulic system is completely sealed. As forces are increased on the push rod, the fluid ahead of the primary cup continues to build up in pressure and begins to flow around the residual pressure check valve into the hydraulic lines. This pressure is then transmitted to the individual wheels, actuating the brakes. When the brake pedal is released, the return springs on the brake shoes forces the fluid back to the hydraulic lines, around the residual check valve and into the cylinder bore. The piston is forced back toward the boot end of the master cylinder, opening up the compensating port allowing fluid to return to the reservoir. Drawing courtesy of Grey Rock.

New master cylinders should be disassembled and inspected, and lubricated with clean brake fluid. There's no need to leave anything to chance. The master cylinder should also be part of a frequent maintenance check, with a tear-down and inspection and rebuild if necessary. Be sure to check the condition of the piston seal.

DUAL MASTER CYLINDERS

While dual master cylinders are standard safety equipment on

The Ford single reservoir master cylinder mounted on a firewall. Be careful in planning the mounting position. We've seen cars where the roll cage tubing had to be notched to make room for the master cylinder. Plan everything out first.

almost all passenger automobiles now, they are still not a good idea for racing applications (with a drum system). A single piston master cylinder insures the same force distribution throughout the entire life of the brake linings. With a dual master cylinder, the pedal is putting equal force on both pistons. If the front linings wear more than the rear linings during a race, the dual master cylinder is automatically then going to deliver more force to the rear brakes. In other words, there is no control over force distribution.

WHICH MASTER CYLINDERS TO USE

As is explained in the chapter on how the braking system works, the best master cylinder to use is a single piston with a one-inch bore. The source of this master cylinder is:

Auto source	*Cyl. bore*	*EIS part #*	*Ford part #*
'65-'66 Mustang	1"	E32900	C2AZ-2140-B
'60-'65 Mercury	1"	E32900	C2AZ-2140-B
'60-'65 Ford	1"	E32900	C2AZ-2140-B
'60-'65 Falcon	1"	E32900	C2AZ-2140-B
'60-'65 Comet	1"	E32900	C2AZ-2140-B

With a disc brake system in a stock car, most disc brake system manufacturers recommend a 1.125-inch bore master cylinder. This is because they assume you are going to experience rigidity problems in the system. The larger diameter master cylinder will supply enough fluid to fill up all the cavities created when the system expands. With a very stiff braking system (with a hard pedal) use a smaller diameter master cylinder to reduce the pedal force required to stop the car. A good 1.125-inch master cylinder is a Chrysler application, Grey Rock part number 36054.

When system rigidity is a problem, some people suggest adding a brake booster to the system to increase pressure. Don't! They are unreliable in a race car—they can break down, resulting in greatly reduced braking efficiency.

BRAKE FLUIDS

Brake fluid is the precious bodily fluid of the brake system. It is the brake fluid which transmits force through pressure from the driver's foot to the brake lining materials. The basic premise which allows this to happen is that fluid is not compressible. Thus, it is able to transmit force.

The enemy of brake fluid, just like friction materials, is heat. If the fluid boils at any point in the system, or if the system leaks at any point, the fluid incompressibility is lost, and thus pedal travel is increased.

The important thing to keep in mind when purchasing brake fluids is that they are not all the same. The main ingredient is ethylene glycol, which has a lubricating capability for the system's rubber parts and is not highly susceptible to boiling.

Brake fluids in the United States are rated by the Department of Transportation. Those fluids used in disc brake systems are rated DOT 3 or DOT 4 and conform to DOT specification 70R3 and SAE specification J1703. The most important characteristic of any brake fluid for race car applications is that of its boiling point. A high boiling point is required for racing applications due to the large heat build-up in the rotor, which can be transferred through the caliper and the pistons into the fluid itself. Most brake bluids manufactured for highway use in the United States have low moisture absorption characteristics. Passenger car brake fluids will absorb moisture much too quickly if additives were not intoduced into the fluid in order to cut down on the moisture absorption characteristics. However, when this occurs the boiling point is reduced significantly. This tends to reduce the usefulness of most DOT specification brake fluids, particularly in severe racing applications. Although moisture will also lower the maximum boiling point of a fluid, the maximum boiling point is an important parameter for selecting a brake fluid for use in a race car. The minimum acceptable dry boiling point is 500 degrees, and preferably higher. The best brake fluid available for race car brake systems is the AP racing Lockheed 550 fluid. There are other excellent brake fluids but they are not able to offer the maximum dry boiling point that the AP does.

It is imperative that the disc brake system be bled before each event. Many top racing teams will replace all of the fluid in the braking system before each race. This is the only way to insure maximum brake performance by keeping the brake fluid from boiling. Also because of the moisture absorption tendency of brake fluid it is a good idea to buy brake fluid in small quantities and to use small containers such as pint cans. The fluid in both the can and in the master cylinders should be exposed to the atmosphere for the shortest possible time. The container in which the fluid is stored should be kept tight and well sealed. It is also a good idea to keep the master cylinder full or to be sure that the master cylinder cover has an expandable rubber boot, which will take up air space as the fluid level decreases.

SILICONE FLUIDS

The newest development in high temperature brake fluids is the silicone brake fluid. It has a much higher boiling point than the traditional ethylene glycol fluid, but it also has some inherent drawbacks.

Many teams have used silicone brake fluids successfully where other teams have encountered problems. Most of the early problems associated with silicone brake fluids have been resolved, particularly those associated with fluid viscosity. However, there are some minor drawbacks to using silicone fluids that the racer should be aware of. First of all, a silicone fluid is more compressible than the glycol based brake fluids. The use of silicone fluids requies a larger master cylinder capacity. If the master cylinder capacity is not adequate, excessive pedal travel will be experienced with the silicone fluid. This problem can be compounded by other problems within the braking system such as excessive rotor run out, pad knockback or inadequately torqued wheel bearings. Another drawback to silicone fluids is related to their better metal-to-rubber lubrication qualities.

Two good silicone brake fluids are GE SF900 and one made by Dow Corning. Both have a boiling point in excess of 700 degrees F. Another excellent silicone brake fluid is made by Yankee Silicones.

Drum Braking System

The drum brake system has been the standard of the industry for years in auto racing, and as a consequence, the drum brake has been developed to a high art.

The needs for stopping 3,800-pound Grand National racing cars at Riverside and Martinsville racing facilities (the two hardest tracks on the Grand National circuit on brakes) have led brake engineers to develop extremely tough, durable, long lasting drum braking systems. The two major areas of concentration for these designs has been in the areas of drums and friction materials.

Before we get into how and why these drum brakes can last through such grueling punishment, let's first examine how the drum brake works.

DRUM BRAKE THEORY

The basic principle of drum braking systems, or any braking system for that matter, is converting motion into heat. This is accomplished through friction. Friction is induced by expanding the brake shoes against a revolving drum. To help induce this friction, one of the biggest advances in drum brake technology is the self-energizing brake.

In a self-energizing brake arrangement, there are two shoes, a

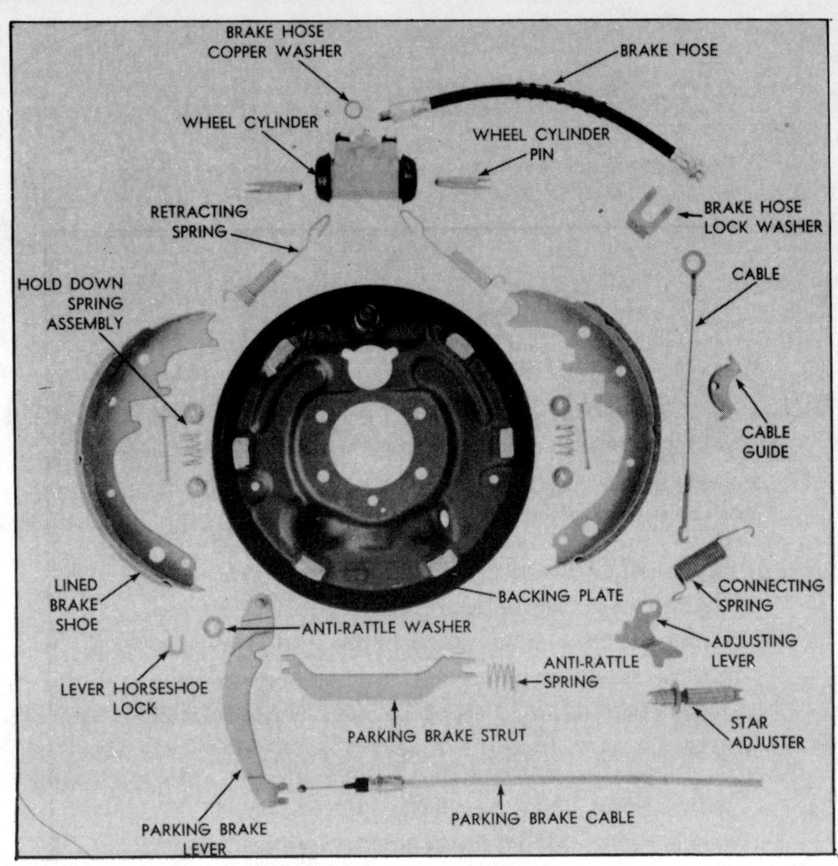

Exploded view of a typical passenger car drum braking system [rear wheel]. In a racing vehicle, the parts relating to the parking brake are eliminated.

primary or leading shoe and a secondary or trailing shoe. As you may be able to guess from the name, in a self-energizing system, one shoe is used to help create more pressure on the other. It works like this: when the driver steps on the brake pedal, pressure is created from the master cylinder through the brake lines to the wheel cylinder where it creates a force against the brake shoes. As the leading shoe makes a firm contact with the revolving drum, the movement of the drum pulls it with it (forcing it to rotate with the drum). This movement is transferred by a link from the primary shoe to the trailing shoe, which forces the trailing shoe much tighter against the drum. The secondary or trailing shoe does not follow the revolving motion of the drum because it is stopped at its top point by a metal anchor pin.

It is the trailing shoe which actually does the largest percentage of the braking work, and thus this is the reason linings are thicker for the trailing shoes. A disadvantage to self-energizing drum brakes is their non-linear response to pedal force (meaning that the force the driver applies to the pedal is not proportional to the force the lining places against the drum). This means the driver has to learn how to modulate the brakes so he does not create a wheel lock-up situation. Of course, with a drum brake system self-energizing is a must. The driver would never be able to apply the force required to stop a stock car.

DRUM BRAKE PROBLEMS

Heat, and the dissipation of it, is the greatest problem concerning the braking system. The harder the brakes work, the greater the heat they produce from converting motion. If this heat cannot be carried away fast enough, brake fade, or impairment of braking ability, is the result. It actually means that the brake shoe friction

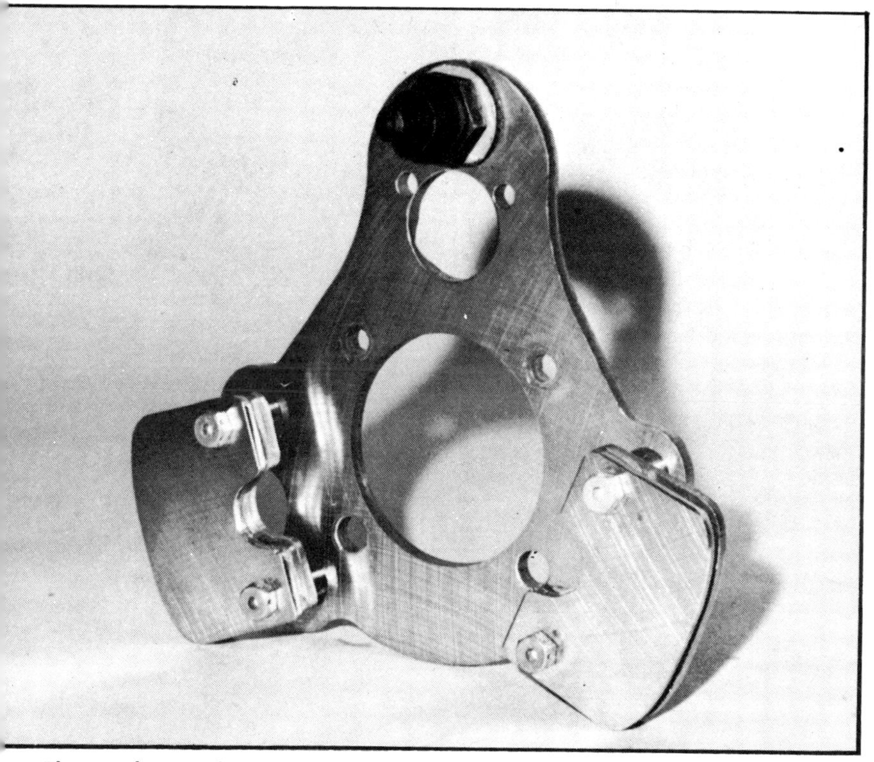

The Speedway Engineering open air brake backing plate.

materials lose their coefficient of friction, and no friction means no stopping.

The ultimate design, then, is one which keeps the brakes as cool as possible. Engineers solved this by creating the open backing plate systems (see accompanying photos of Speedway Engineering's Open Air backing plate system). This design incorporates as little as possible in the way of backing plate materials and components, so there is more air flow to the brake shoes and drums. A side benefit with a specially-constructed component such as this is that areas which should be reinforced or constructed more heavily can be, such as incorporating a heavier

Three common brake drum problems illustrated here are hard spots on the drum surface [right], bell-mouthed shaping of the drum [lower left] and scored drum surface [lower right]. Drawings courtesy of Grey Rock.

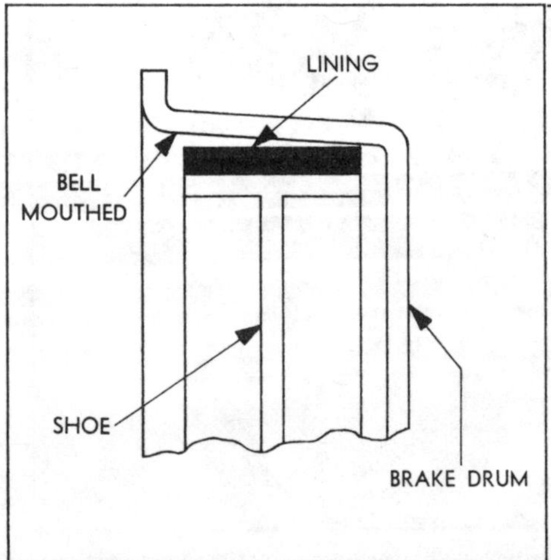

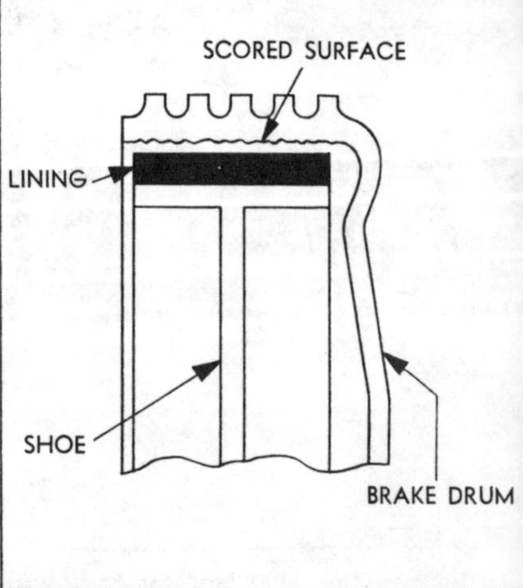

anchor pin and stiffer return springs (these springs can easily lose their strength with the high heat generated).

Heat is also a great enemy of drums. When metal gets hot, it can lose some of its strength and rigidity. With brake shoes creating a great force against the drums at a time like this, it is easy for the drums to lose their shape. Drums elongate themselves, or grow "bell-mouthed," permanently. When this happens, brake shoes cannot exert a steady, even pressure against the drums and some loss of braking ability is suffered. Another problem with drum brakes is as they heat up they temporarily get much larger, and therefore more pedal travel is required to use up the clearance. The same thing happens as the lining wears. If a self-adjusting feature is incorporated, such as disc brakes automatically have, the drum brake may drag or lock up if the drum cools off and contracts.

BRAKE DRUMS

When it comes to selecting brake drums which are up to the job of stopping a stock car, there are very few good ones available. Back in "the old days," extremely heavy cars such as the Lincoln Continental and the Chrysler Imperial were stopped by drum systems, and thus faced some of the fade and reliability problems that the stock cars do today. But, the braking systems in these cars have now been replaced by disc systems and the once plentiful finned 11" by 3½" all cast drums are hard to find. For a list of drums which are available, see the accompanying list.

For heavy duty braking requirements, 3½" wide brakes in the front and 3" wide brakes in the rear must be used. On tracks where braking requirements are not so severe, 3" wide brakes in the front and 2½" wide brakes in the rear may be used.

Two of the Century replacement drums.

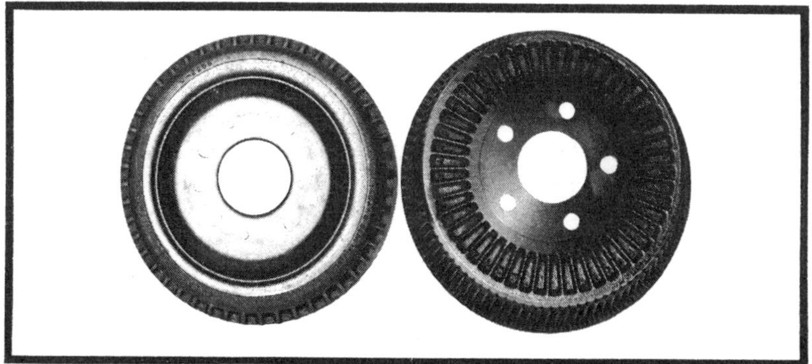

BRAKE DRUM SOURCES

Part #	Mfr.	Size	Wheel Pattern	Note
2268	Century (9)	11 x 3	5 x 5	1, 2, 8
2816	Century	11 x 3	5 x 5	2
1280	Century	11 x 3	5 x 5	1, 3, 4, 8
1270	Century	11 x 3	5 x 5	1, 4, 8
2065	Century	11 x 3	5 x 5	8
2923	Century	11 x 3	5 x 4½*	3, 5
2959	Century	11 x 3	5 x 5	3
2534905	Chrysler	11 x 3	5 x 5	3
3461715	Chrysler	11 x 3	5 x 5	3
3580567	Chrysler	11 x 3	5 x 5	3
3412097	Chrysler	11 x 3	5 x 5	6
3690599	Chrysler	11 x 3½	5 x 5	7
'58-'60	Lincoln	11 x 3½	5 x 5	

* A machine shop can redrill pattern to 5 x 5.

Notes

(1) Has 3-1/16" register for centering on hub
(2) Thin metal center
(3) Can be machined to 3½"
(4) All cast
(5) For short track racing only
(6) Very heavy duty drum
(7) Now obsolete
(8) Finned
(9) Century brake drums are available from Grey Rock or Raybestos Manhattan brake hardware suppliers.

BRAKE COOLING

Forced brake cooling helps dissipate heat to keep braking efficiency up and temperatures down. If you are going to compete in a long distance race on a short track, such as 300-500 laps on a 3/8 or 1/2-mile track, you should use flexible plastic ducts to force the cooling air to the brakes. Duct the air from a high pressure area in the front of the car, such as the parking light openings or in the spoilers. Use a minimum of 3-inch inside diameter cooling ducts to supply a sufficient amount of air. Even a larger diameter ducting is better.

If your car is drum brake equipped, duct the air through the backing plates. With a vented disc system, direct the air to the center of the disc. With a solid disc (which should not be even

The Chrysler 3690599 drum is shown above. Below is the left front wheel of Richard Petty's Dodge set up for Riverside. Notice ducting of air into the backing plate through 4" diameter flexible tubing. Notice also how it is taped to the upper A-arm to allow it to move as part of the suspension system.

Air flow ducting into a disc brake set-up.

considered for a stock car racing application), the air should be directed to the caliper.

After mounting the brake ducting, check to be sure it does not get hung up on the wheels or suspension when the wheels are in full lock or the suspension is in full bump or rebound.

Be sure to use a coarse screen over the ducting opening to prevent introducing rocks and small gravel into the system.

BRAKE PROPORTIONING

Achieving a perfect balance between the front wheel and rear wheel braking effort under heavy braking can be a major problem. Each wheel should have braking force applied to it in direct proportion to its ability to transmit that force to the ground.

If too much braking effort is applied to one pair of wheels, the other pair is not doing enough of the work, and the result is that the most heavily biased pair will lock up. With a locked-up front pair of wheels, heavy understeer is created which will last until the brakes are unlocked. If the rear wheels lock up, the car is most generally going to spin out. The optimum and most stable braking condition is, of course, to have the front wheels lock up just an instant before the rear wheels.

Major brake bias to the front is adjusted by fitting larger wheel cylinders at the front, with a consequent increase in pedal travel. Fine tuning of the braking system can be done with a manually adjustable brake proportioning valve. Front braking bias usually is found on most racing vehicles to be between 56% and 80%.

Many conditions will vary the amount of front-to-rear bias required. These conditions include track shape, smoothness, and coefficient of friction, tire section size, tire compound, vehicle speed, vehicle weight distribution, as well as varying driving techniques. •

COMPUTING BRAKE FORCE DISTRIBUTION

The exact braking effort distribution partially depends on the length of the wheelbase and the height of the center of gravity (CGH) of the car. The reason for this is the load transfer caused by these two elements during braking which pitches weight forward. The more load put on a tire, the lower its coefficient of friction, so braking performance is definitely affected.

To illustrate how weight affects braking performance, let's say you have a car which can stop at 1 G (a reasonable goal for racing vehicles) and weighs 3,000 pounds. Suppose you were able to lower the weight to 2,000 pounds, and all other elements remained the same. With this change, the car will probably be able to generate anywhere between 1.025 and 1.1 G of stopping force (the variable depends on tire compound and construction).

As an example of the effect of load transfer on stopping ability, let's say you have a car weighing 3,000 pounds, has a CGH of 18", a wheelbase of 100" and 50/50 static weight distribution. When you hit the brakes on this car and try to stop at 1 G, some of the static load on the rear wheels will be transferred to the front. To compute how much is transferred, multiply the CGH times the weight of the car times the stopping G, and divide all of this by the wheelbase. In our example, we would have 18" x 1 G x 3,000 lbs., divided by 100". The answer is 540 pounds, which means that when stopping at 1 G, 540 pounds are subtracted from the rear tires and added to the front tires. This same amount of weight transfer will occur no matter what the static weight distribution is. Since the front and rear tires started with 1,500 pounds each (per pair), the front now has 2,040 pounds and the rear has only 960 pounds. This means that, with equal size tires front and rear, at least 68 percent of the braking effort will be required on the front wheels (2,040 is 68 percent of 3,000). Furthermore, the coefficient of friction of the front tires will be reduced between .01 and .05, and the rear tires will be increased by the same amount. Assuming that there is a change in coefficient of friction of .03, the braking force generated by the front tires will be .97 times 2,040, or 1,980 pounds of braking force. At the rear, we will have 1.03 times 960, or 999 pounds of force. The total braking force will be 1,980 plus 999, or a total of 2,979. If we take this total braking force and divide it by the total weight of the vehicle, we get only .99 G of braking deceleration instead of 1 G.

The best braking set-up for any car is to have about 70 percent of the static weight on the rear wheels (which is difficult to achieve). 70 percent of weight on the rear wheels is also good for low speed acceleration, needless to say.

There is an important aspect to braking force distribution for cars which use a front spoiler. When any car (with or without a lot of anti-dive built into its front suspension) brakes hard, it pitches forward. This pitch brings the front spoiler much closer to the ground, cutting the airflow under the front of the car to almost nothing. When this happens, the aerodynamic downforce is greatly increased. It would not be unusual for this increase in downforce to exceed 500 pounds at high speed. This increase will prevent the front wheels from locking before the rear wheels, and therefore causing the car to spin out if the rear wheels are locked up. The only thing which can be done about this problem (short of putting a lot of undesirable anti-dive in the front end) is to set the brake balance for the highest speed corner on the race track.

THE BRAKE PROPORTIONING VALVE

The manually adjustable brake proportioning valve was originally developed to proportion braking effect between disc front brakes and drum rear brakes all being fed from a single master cylinder. With this type of braking setup, disc brakes require much more hydraulic pressure to operate than do the drum brakes, so consequently the rear drum brakes are being fed too much pressure and are locking up. To combat this, the brake proportioning valve is installed in the line between the master cylinder and the rear brakes. During normal stops, the valve does nothing, but under hard braking the valve comes into operation around 200 pounds per square inch of line pressure. When adjusted, it only allows the line pressure to the rear wheels to increase as a percentage of the line pressure being fed to the front brakes. This percentage is what is manually adjustable.

Manually adjustable brake proportioning valves are available from Chevrolet (PN 3878944), or from Tilton Engineering.

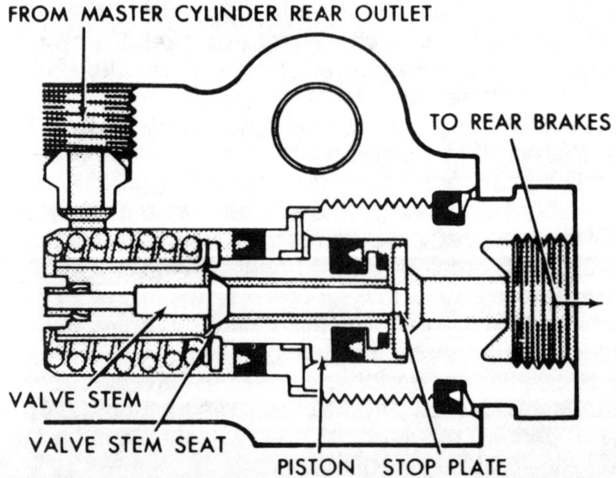

How the brake proportioning valve works: on a hard brake application, pressure pushes against the large end of the piston. When that pressure is sufficient to overcome the spring load, it moves the piston to the left so pressure flow through the valve is restricted. The spring load is what is adjusted on the valve.

The manually adjustable brake proportioning valve available from Chevrolet, PN 3878944.

TESTING BRAKE BIAS

The best way to test braking bias is to brake hard from about 75 miles per hour on a dry paved surface. Have an informed observer nearby who can tell which pair of wheels lock first. Continue to adjust the brake proportioning valve until the front wheels are barely locking up first. Be sure you are using the same type tires during the testing session which you will be racing with.

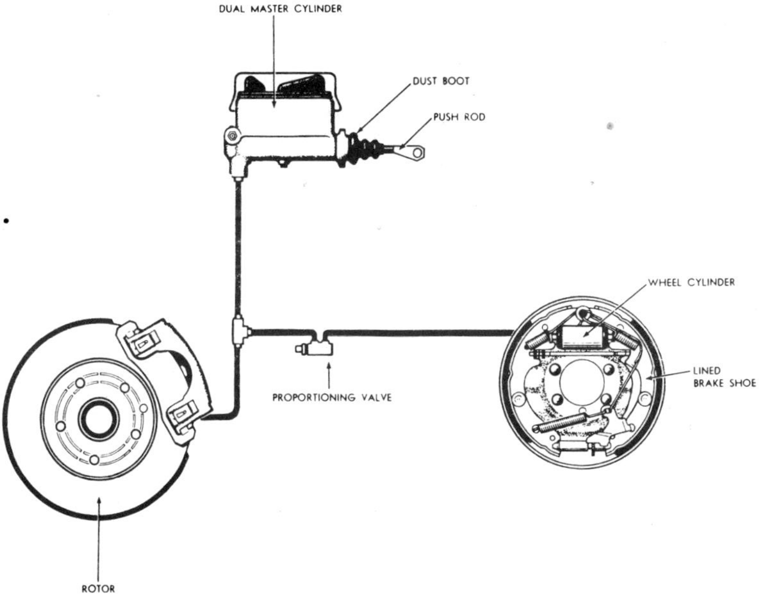

DUAL MASTER CYLINDER
DUST BOOT
PUSH ROD
WHEEL CYLINDER
LINED BRAKE SHOE
PROPORTIONING VALVE
ROTOR

Schematic drawing of how the proportioning valve should be installed in the braking system to prevent rear wheel lock-up. This would be true even if front wheel was a drum braking system and master cylinder was a single reservoir type.

Disc Braking System

In the past few years there has been a rapid switch by passenger automobile manufacturers from the drum brake system to disc braking systems. There are four basic reasons why this has taken place: (1) Disc brakes have a better ability to resist heat and moisture-induced fade, (2) Disc brakes can afford a weight saving, (3) Disc brakes have a simple design affording ease of maintenance, (4) Disc brakes have a linear response to pedal force. In addition, disc brakes have a built-in self-adjusting feature when the brakes get hot and when linings wear. The rotor expands as it gets hot toward the friction material and not away from it as a drum brake would.

For these same basic reasons, there is now a change in high performance racing vehicles from drum to disc systems. The change did not take place as rapidly in racing vehicles as in passenger autos, however, because of two problems with disc systems in racing: (1) A rapid enough heat dissipation that would not boil brake fluids, and (2) The availability of high performance disc brake pads. These problems are now resolved, and there are several different approaches to the disc system which are available for the race car. Before we discuss these, however, let's investigate the basics of the disc system.

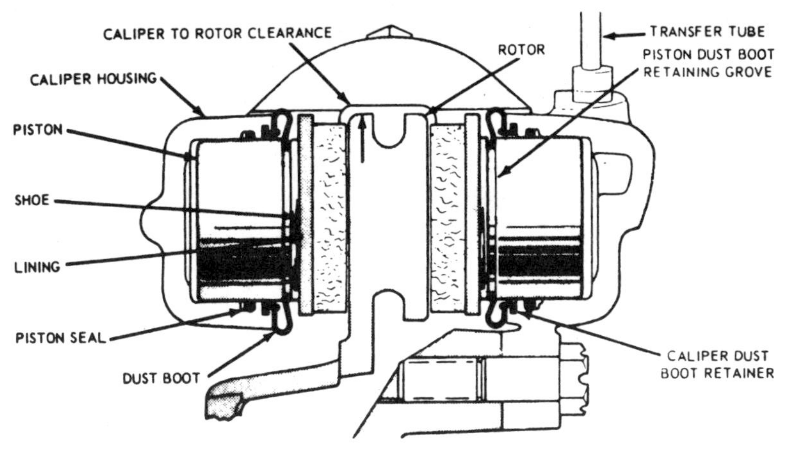

Typical terminology of the disc brake system components.

DISC SYSTEM BASICS

Disc brakes are not self-energizing which means that the driver must apply a continual hard pedal effort directly proportional to the stopping effort. This also means that there is only pure hydraulic pressure forcing the pad against the disc, so disc systems require much greater pad force than do drum systems. This can be accomplished through increasing the lever arm ratio of the brake pedal arm, by increasing wheel cylinder sizes, or by decreasing master cylinder bore sizes.

There are four criteria which are vitally important when choosing, designing and working with a disc brake system: (1) Keep caliper deflection down, (2) Use hard linings (to avoid flex from sponginess), (3) Use small diameter flex lines, and (4) Use steel brake lines wherever possible.

The reason behind all of these criteria is the elimination of deflection in the system. Deflection can be described as the bending or flexing of a component when loads are applied to it, and with the extremely high pressures and forces required in a race car disc system, deflection can be a large headache.

Deflection, really, is what can make or break the abilities of a disc system. The less deflection, the better the system will work. For example, if the flex lines are of rubber instead of stainless steel braided teflon, pedal travel and fluid volume will be lost by expanding the diameter and length of the flex line. If the caliper housing flexes, force intended to be used in pressing the lining against the rotor will instead be used to deflect the housing and there will be little force left for stopping the vehicle.

THE CALIPER HOUSING

The first item to consider in selecting a good disc brake system is the caliper housing deflection. To explain why this deflection is so detrimental to the caliper, refer to the accompanying drawing. Remember that basic law of physics, "For every action, there is an opposite and equal reaction?" Well, it certainly applies in the caliper housing. In the drawing, "A" refers to the initial action of the piston being forced against the pad, which in turn is forced against the disc. "B," then, represents the opposite and equal reaction. If the caliper housing were flexible enough, the system could use the rotor as the anchoring or back up point, and the fluid pressure would then apply force against the housing. This would deflect the caliper using all available pedal travel and deliver little if any stopping ability.

There are three basic items in the design of a caliper which determine how much it is going to flex: the material from which it is made, how thick the material is, and the bridge distance (or distance between clamping bolts).

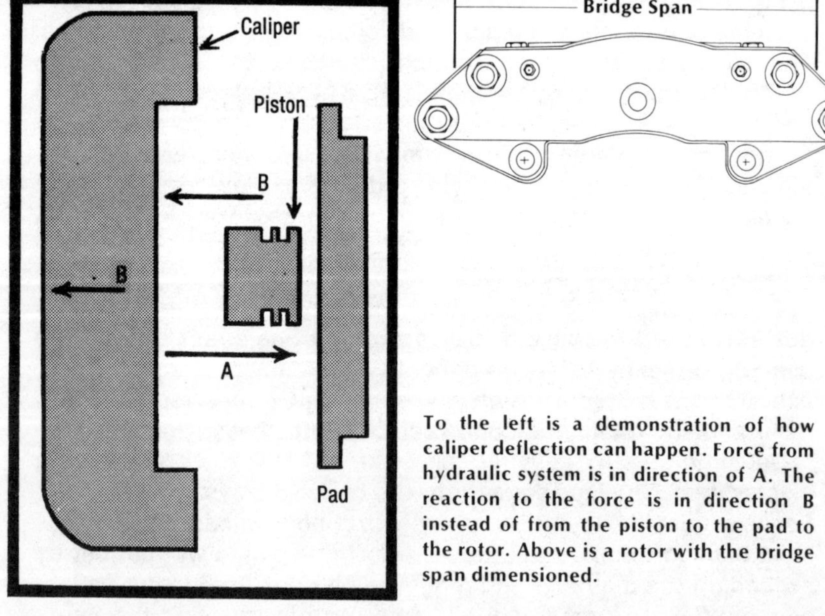

To the left is a demonstration of how caliper deflection can happen. Force from hydraulic system is in direction of A. The reaction to the force is in direction B instead of from the piston to the pad to the rotor. Above is a rotor with the bridge span dimensioned.

Above, an illustration of how caliper deflection can be determined. A hydraulic cylinder delivers a measured amount of pressure into the caliper, and the deflection of the brake caliper can actually be determined with a measuring caliper. 3,000 p.s.i. was delivered to the JP caliper in this photo, with virtually no deflection. Most other calipers measured showed serious problems past 1,200 p.s.i.

Below, this is a modified we photographed at Daytona which had rubber flex lines leading to the calipers. Talk about asking for problems!

Calipers are generally made from three common casting metals: magnesium, aluminum and cast iron. The common denominator of all these metals, in relationship to their flex, is their modulus of elasticity. That is a number which relates the metals in their resistance to being deflected. The higher the modulus of elasticity number, the greater the resistance to flex. Magnesium has a modulus of 6.5 million. Aluminum has a modulus of 10 million. Cast iron has a modulus of 14.5 million. And when we talk about floating calipers a little later, we will want to know that the modulus of steel is 30 million. So we can see from these figures that magnesium, while it is good for such lightly-stressed castings as rear end housings, is not particularly preferable for an efficient disc braking system. Magnesium's greatest asset is its low weight, but for a caliper it would be better to sacrifice a little weight to reduce caliper flex.

One of the biggest causes of deflection in a caliper is the use of inadequately sized bridge bolts. This is especially true of many aftermarket calipers of the two-piece design. The largest possible bolts should always be used in the area of the bridge to minimize deflection. Only aircraft quality, AN bolts and nuts should be used. It is important that the holes be reamed for proper fit and that the proper grip length bolts be utilized to eliminate wear and sloppy fits.

Most aftermarket calipers use an open back caliper construction for quick pad removal. In order to minimize caliper deflection, this type of caliper must be made much larger than is actually necessary. Most race cars do not need quick change calipers and significant weight savings can be realized when this feature is eliminated. In addition, most aftermarket calipers are sand cast from 356 T6 aluminum. A pressure casting of alloy aluminum will provide greater strength per unit of weight.

The bridge distance is the span between clamping bolts. The longer this span, the thicker the caliper housing material must be to resist deflection. Or, the longer this distance, the greater the deflection is going to be. So, the piston or pistons should be placed as close to the caliper clamping bolts as possible.

The question of bridge span is closely related to the piston size and arrangement. The simplest piston arrangement is the single piston on one side of the rotor. Then there is a single piston on each side of the rotor, two pistons on each side of the rotor and three pistons on each side of the rotor.

The best arrangement of pistons is two or three in a row on each side of the rotor. This spreads the force loading closer to the clamping bolts, putting it in combined tension and bending against the bolts, reducing caliper deflection.

PISTONS AND "O" RINGS

OEM brake calipers normally utilize steel pistons with a dust deal and a small 'O' ring seal. Aftermarket calipers are generally made of steel with a black oxide outer coating. Aftermarket calipers generally do not have dirt seals and thus are subjected to increased wear from dirt, calcium and water. To minimize the possibility of sticking pistons due to wear, racing pistons should be made either from chrome plated steel, anvilized aluminum, stainless steel or titanium. Pistons of these materials will resist corrosion and sticking. 'O' ring assembly lube should be used between the seal and the outside of the caliper when calipers are assembled. This helps to exclude dirt and moisture from the piston and piston bore.

Most racing calipers utilize a round section 'O' ring for sealing. The 'O' ring serves two functions — first, to seal the piston within the bore and stop fluid from leaking to the outside of the caliper. Second, the 'O' ring helps to retract the piston when fluid pressure is released. Round section 'O' rings tend to create problems, however. If the cross section of the 'O' ring is too small, the pistons can be easily knocked back by external forces and momentary pressure reductions. If the cross section is too large, the 'O' ring will hold the pad against the rotor when pressure is released, thereby causing brake drag. The proper seal shape for a disc brake system is a square section 'O' ring. A square section 'O' ring will fill the entire groove, and thereby reduce loss of pedal pressure. It also provides a full width sealing surface and will automatically retract the piston when hydraulic pressure is released.

HOW MUCH PEDAL TRAVEL?

Disc brake systems are set up with the pads riding about .002" away from the rotor, so many people assume that very little pedal travel is required to actuate the system. This is not so. Let's assume we have a front wheel disc brake system which employs eight pistons, each with a two-inch cylinder bore. Each wheel cylinder travels .006" (taking up the clearance plus allowing for pad material compressibility). The volume of fluid required for this wheel cylinder to travel .006" is .006" times the area of the cylinder (which is pi times the radius squared) which works out to .019 cubic inches. Multiply .019 cubic inches by eight pistons and we have a fluid movement of .152 cubic inches. How much piston travel (in inches) does this fluid displacement mean in the master cylinder? Assume we have a 1" bore cylinder, which is .79 square inch of area. So, .152 square inch equals .79 square inch times x, or

.192" of travel, of the master cylinder piston. If the pedal leverage is 4 to 1, then this translates to a little more than 3/4" (4 times .192 equals .768 inch) of pedal travel. We must add to this pedal travel the pedal deflection and master cylinder deflection (assume 1/2"), and caliper deflection (assume .003 inch per caliper under 500 pounds of force, which works out to another 1½" of pedal travel). So, to just begin applying force to the front rotors from the piston and lining, we have used up 2-3/4" of pedal travel! This, of course, is assuming very low amounts of deflection throughout the rest of the system. Can you imagine what kind of pedal travel would be required with a high deflection system?

MOUNTING THE CALIPER

When it comes to mounting the caliper onto your present suspension system, you will have to fabricate your own mounting brackets, whether you purchase a specialty stock car product such as the Edco system or the Hurst/Airheart system, or adapt a passenger car application to your car.

For the brake system to function properly it is important that calipers be mounted to the spindles as rigidly as possible. In many instances disc brakes are installed on spindles that were originally intended for drum brake applications. This requires the fabrication of a special mounting bracket. In most instances the mounting brackets are made of inadequate material, usually 3/16" thick steel which is inadequate. This is especially true of the heavier sedans and stock cars. The brackets should be a minimum of 1/2" steel or 5/8" aluminum and every attempt should be made to mount the brackets in a single plane relative to the spindle. One way to utilize a lightweight bracket and to improve the strength of the mounts is to use two 3/16" thick steel brackets and mount the caliper in double shear by sandwiching the ears of the caliper between the two brackets.

Another design parameter in mounting the caliper is to mount it so that it is to the back of the rotor. It is improper to mount a caliper at the front of the rotor becausee its effectiveness is diminished. When the caliper is mounted in front it increases spindle deflection. If the caliper is mounted at the top or bottom of the rotor, maximum spindle deflection will be transferred to the caliper resulting in a loss of braking efficiency. It is also difficult to mount the bleed screw at the highest point in the system if the caliper is mounted at the top or bottom of the rotor. Rear mounted calipers will also allow improved air flow or easier ducting of cool air to the brake rotor.

When calipers are mated to solid axles or to chassis on inboard suspension systems, the overriding design parameter is the location of the caliper bleed screw.

FLOATING VS. NON-FLOATING CALIPERS

The non-floating caliper is one which has pistons on each side of the rotor, squeezing or pinching the rotor from each side. Because hydraulic pressure is always self-equalizing throughout an entire system, one piston cannot overcome any others. Nothing moves in the caliper except the pistons pushing the pads inward from each side.

The floating caliper was designed by passenger car manufacturers essentially to make the caliper less expensive to produce. It successfully applies the physics principle of "every action causes an opposite and equal reaction." Applying the principle, they eliminate pistons on one side of the rotor. The full-floating caliper is not solidly mounted but rather slides back and forth slightly on bushings. When braking force is applied, the piston pushes the pad on the primary side. Instead of the opposite and equal reaction, being supported by a rigid caliper housing, it pushes against a stamped steel or cast iron caliper. This caliper, which basically is shaped like a horseshoe, has a brake pad on the other side of the rotor and is pulled tight against the rotor.

This caliper must be very rigid, of low deflection, or the entire principle behind it is lost, and so is braking ability. This is why full-floating calipers are made of steel or cast iron—it offers a high modulus of elasticity.

The full-floating caliper can offer one big advantage. If the rotor has a slight run-out (or has a wobble), the floating feature will compensate without creating any instability. But with a non-floating caliper, this rotor wobble would create quite a bit of instability and pad knockback under hard braking. Another advantage of the single piston caliper is that there is just one piston to worry about bleeding. The bleeding of air through the tiny passages which connect multiple pistons can sometimes be maddening and next to impossible.

Be sure to check the condition of the nylon bushings which the caliper floats on if you use the full-floating system. If the sliding movement binds up on the bolt, there will not be full force application against the rotor.

THE ROTOR

A rotor is not just a rotor. Each different disc brake system has its own rotor, each with its advantages and disadvantages. Basically, they vary in size both in width and in diameter.

Disc brake rotors fall into two categories: Solid Rotors, and Ventilated Rotors. As the name implies the solid rotor is a solid piece of material. The ventilated rotor on the other hand, is a cast piece with braking surface on the outboard sides of the rotor and fins in the center. The fins not only reduce the weight of the rotor but help dissipate heat and improve air flow which helps cool the discs.

The primary design parameter for any disc brake system is the ability of the system to dissipate heat. It is the limiting factor of any disc system — and the ability of the caliper to generate a clamping force. The temperature of the rotors and thereby the entire braking system is critical to maximum braking performance. If the rotors are not able to dissipate heat quickly enough, the brake fluid can boild and total system failure can result. Monitoring the temperature of the brake system is important. There are various types of temperature monitoring devices which can be utilized to determine the temperature of the braking system at various points. The most notable are the Tempil Stick, which is a temperature indicating crayon; the Tempilaq, which is a temperature indicating liquid; and Tempil Label, which is an adhesive backed stick on a temperature indicating device. Both the crayon and the liquid will melt at a specific temperature. The ring on the Tempil Label will turn black when the indicated temperature is reached. These devices help tremendously in determining the efficiency of the braking system.

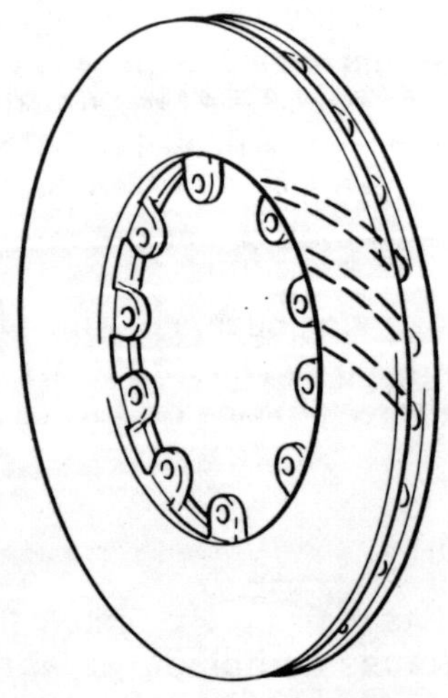

A number of manufacturers have solved this problem by making rotors with curved ribs or veins. The curved ribs allow the walls of the rotor to be completely supported at all times and thereby, minimize deflection of the rotor. The curved veins also provide a larger surface area for cooling and improve air flow within the rotor.

This is a Corvette front wheel disc system being employed on an IMSA road racing Corvette. Notice the air duct

The choice of rotors in the disc brake system, either solid or vented, is primarily a function of the weight of the vehicle, the speed of the vehicle and the amount or frequency of deceleration. on a lightweight vehicle, solid rotors are used almost exclusively. Vented rotors are always used at the front of heavy vehicles when frequent and extreme deceleration is required. Generally any car which is able to run a lightweight rotor would usually be able to run a solid rotor which in most cases is lighter and has reduced distortion and increased resistance to cracking. The manufacturer of any brake system should be able to tell you what is required to achieve maximum efficiency for your particular racing vehicle. For heavy vehicles requiring numerous high speed decelerations, the largest diameter and widest vented rotor is the most desirable.

Most rotors have straight ribs radiating from the center to the outside of the rotor. They have to be placed far enough apart so air can flow between them for cooling. But this leaves areas of the rotor wall which fall between the ribs which have no backing support. When the rotor is squeezed by as much as six tons of force, this causes rotor wall deflection between the ribs, even resulting in cracking. The problem with rotor wall deflection is less uniform pad surface pressure with the rotor, thus less stopping ability.

ROTOR CRACKING

Keeping an even temperature across the entire rotor from center to outer circumference is an immense problem with disc brake systems because of non-uniform heat application to it.

This problem is particularly severe with the one-piece cast rotor, hat and hub which is universally common to most all passenger car disc systems. The problem is that heat is applied to the outer circumference of the circle, and this part is expanding at a greater rate than the inner portion. Something has to give, so the rotor cracks and/or warps. The way to cure this is to have a rotor which bolts to a hat with slotted holes (which allows the rotor to grow), which in turn bolts to the hub. This is the common design for the racing-only disc systems. If you are using a passanger car disc system with a one-piece hub and rotor, however, check for cracking weekly. Also check the wheel bearings weekly as the cast one-piece hub and rotor transfers a tremendous amount of heat to the wheel bearings.

ROTOR COOLING

On disc brake applications where heat is a problem, it is often necessary to duct cold air directly to the rotor. This is especially true if the normal air flow around the car does not reach the rotor. When using solid rotors it is important to duct air to both sides of the rotor equally so that distortion due to uneven temperatures is not encountered.

When using the vented rotors it is important that air be ducted directly into the eye of the rotor. The centrifugal effect of the rotor as it turns will draw air from the middle toward the outside of the rotor, thereby increasing cooling. In both instances it is important to end the ducting as close to the rotor as is practical. It is also important to allow for clearance between the ducting and suspension components through their full travel.

ROTOR CHECK LIST

When giving your braking system a periodic check, light ridges, rust and light scoring of the rotors can be taken care of by truing the rotor in a lathe. Heavy ridges, bad scoring, and excessive run-out all require replacement of the rotor if the vehicle is to stop when the pedal goes down. Check for run-out by mounting a dial indicator on the end of the spindle and rotating the rotor against it. The maximum tolerance should be .002". When the rotor is machined in a lathe, it should be indicated from the hub bearing surface (the hub should be bolted to the rotor if it is a two-piece assembly). This way you are certain the rotor runs true from the spindle.

A DISC/DRUM SYSTEM

Some people may want to try disc brakes in the front and drum brakes at the rear. For a racing application, we particularly do not recommend it because of the many problems the racer will encounter in making it work correctly and efficiently. But, if this type of system is desired, here are some tips to make it work better.

Use a Chevelle disc-drum master cylinder, part number 5470665. This master cylinder has an increased fluid capacity for the front brakes, and has the residual check valve removed from the front brake line outlet.

Use a manually adjustable brake proportioning valve inserted in the brake line running from the master cylinder to the rear brakes. The biggest problem of adapting the disc front/drum rear system to racing cars is proportioning the rear brakes to prevent them from locking up first. This proportioning valve can help.

The best combination system to try is either the Chrysler "C" body calipers or GM big car group single piston calipers in front with 11" x 3" drum brakes in the rear. The Velvetouch linings in the rear with Grey Rock pads in the front would be the best combination of friction materials with this system.

The Chevelle disc/drum master cylinder. Notice how disc brake master cylinder reservoir has a much larger capacity.

TACKLING SOME DISC BRAKE PROBLEMS

There are several problems which are common when people first install a disc brake system on a race car. We will discuss them here to save many hours of wondering, worrying and cussing.

When a disc brake system or combination disc/drum system is

installed with a drum brake master cylinder, many times the pads will wear very rapidly and the braking ability will fade from rapid and excessive heat build-up. This is caused by not eliminating the residual check valve from the master cylinder. This valve keeps a low amount of line pressure in the system for the drum system. However, in the disc system, when the driver's foot is off the brake pedal, the line pressure should return to zero in order to allow the piston retraction to be handled by the piston seal. If the pistons do not retract completely, the force is not completely relieved from the pads and they will rub against the rotor.

For persons not acquainted with a disc system, the way the pistons retract can raise some questions. There are no return springs at all. Instead, the pistons are held in place by some very elastic seals. When the piston is pushed out of its bore, the seal does not slide on the piston but rather is stretched. As soon as the pressure is relieved from the piston, the seal brings the piston back into its bore. A light spring behind the piston keeps it from being pulled too far into the bore. Usually the piston movement is no more than a few thousandths of an inch.

Because the seal acts as the retractor, it should periodically be checked and replaced in a race car braking system. If the seal gets old and is not doing its job, it could cause the pad to drag on the disc, or leak fluid.

Because hydraulic pressure is self-equalizing to all pistons, a non-floating caliper should be checked and double checked to be sure the rotor is centered in the caliper. If it is not, shim or space the caliper so as to prevent severe pulling, or losing a piston seal when the pads are worn down.

Adjust the front wheel bearings on the tight side, then check for

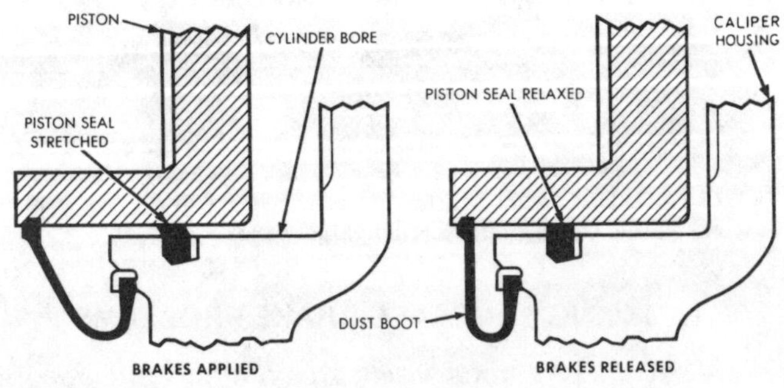

FUNCTION OF PISTON SEAL

Drawing courtesy of Grey Rock.

any rotor wobble. Wobble knocks back the pads and increases pedal travel.

Thorough and frequent brake line bleeding is recommended to eliminate all air from the system and keep it eliminated. Tap the calipers with a rubber or plastic mallet to aid in loosening any air bubbles which may be trapped in tiny passageways.

Make sure that the master cylinder has full travel before the brake pedal contacts the floorboard.

In the event rear brake lockup is encountered under hard braking, install a manually adjustable brake proportioning valve in the brake line leading to the rear brakes.

Only the highest quality, high boiling point brake fluid should be used. On some passenger car disc systems adapted for racing application, there is a problem of heat dissipation into the fluid at the caliper.

Disc braking system on the rear of a Porsche Carrera. The fins on the caliper are to aid caliper stiffness and aid in the dissipation of heat.

HEAT SHIELDING

One of the biggest problems facing designers who have tried to fit disc brakes to heavy racing vehicles is the dissipation of heat into the brake fluid, which naturally is going to cause at least momentary fluid boiling.

Cast iron rotors will heat up to 1100 to 1200 degrees F in race cars. The best high performance brake fluids will boil at 700 degrees F, so this means 400 to 500 degrees F have to be shielded from the fluid.

Heat from the rotor is carried through the pad to the front surface of the caliper pistons. The pistons will conduct the heat right into the fluid. So, it is someplace in this heat transfer process where the heat must be diverted before it has a chance to meet the fluid.

Caliper casting material is very important in heat shielding. Aluminum is a good conductor of heat. Cast iron does not conduct it nearly as well, so this is another plus for having cast iron calipers. Magnesium conducts heat just slightly worse than aluminum.

The first way to tackle the heat problem is having as wide a rotor as possible, affording maximum air flow and air turbulence to transfer heat out of the rotor to the passing air stream. A massive rotor also helps store heat until it can be dissipated. This is the reason the Hurst/Airheart Stock Car Disc Brake employs a rotor 1.375" thick, which is thicker than any passenger car rotor available. You have to look to one-ton truck disc brake systems to find a rotor this thick.

If heat is getting past the pistons and into the fluid (which is very possible with the pads residing .002" away from the rotor at a maximum), then some shielding at the pistons may help. But, when the heat gets this far, it is quite expensive to remedy it.

Chevrolet came up with an insulator shield for the piston which fits right over the front of it. The part number is 5463822 for a 1½" shield which fits the 1-7/8" O.D. pistons, and 5469593 for a 1" insulator which fits the 1-3/8" O.D. pistons.

A very inexpensive, and quite effective, method of shielding heat between the pad and the piston is the use of thin strips of stainless steel. Cut two or three pieces of material from .010-inch sheet stock which covers the entire back area of the pad. Stainless steel is a very poor conductor of heat.

Another effective heat shielder is the new phenolic (a plastic type material) pistons being used in the big-model 1977 Chrysler cars. They conduct very little heat. And, they are large enough that they can be machined to fit many different applications.

DISC BRAKE APPLICATION

In addition to the OEM disc brake systems, a number of companies produce aftermarket disc brake systems for race car applications. These include Edco Specialty Products, Hurst-Airheart, JFZ Engineering, Wilwood Disc Brakes, Speedway Engineering, Winters Performance, Midwest Race Engineering, AP Racing and others. When choosing a disc brake system for a race car it is a good idea to acquire information from a number of the manufacturers. This will allow you to make a better determination of the best product for your specific needs. In the following sections some of the representative aftermarket and OEM brake systems are outlined.

EDCO

The Edco disc brake system recommended for stock cars weighing 2800 pounds to 3900 pounds is their number C-6300-B front calipers, C-4200-B rear calipers, D-1200-U rotors at all four wheels, A-1200-SS rotor to hub adaptors (or hats) and a 7/8" bore master cylinder for the front wheels and a 1" bore master cylinder for the rear wheels.

All Edco calipers are of the non-floating design, cast from

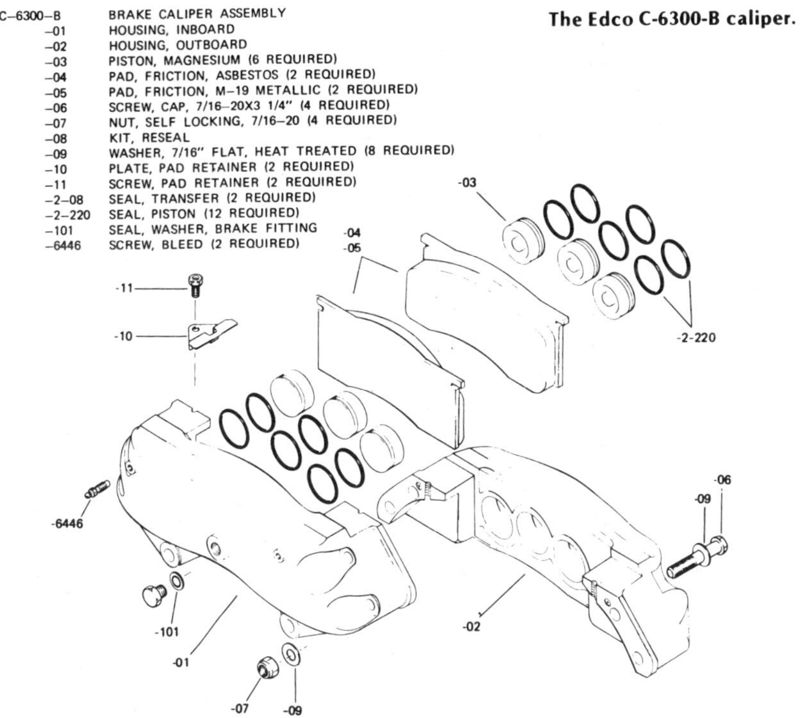

The Edco C-6300-B caliper.

magnesium, but are designed to overcome problems of housing flex by using more pistons of a smaller diameter to put expansion loads in tensile stress in the housing clamping bolts. The front calipers have six pistons each, the rear calipers have four each. The calipers are designed to use standard Grey Rock M-19 material lining pads. The rotor is 7/8" thick. The master cylinders are Edco designed, cast and machined, based on Girling aluminum units.

When mounting the Edco master cylinders together in tandem for use with a balance bar (which balances driver force application to the front and rear cylinders), be sure that the mounting bracket is as rigid as possible. Do not try to save weight here because stout and sturdy materials will be good insurance against force losses.

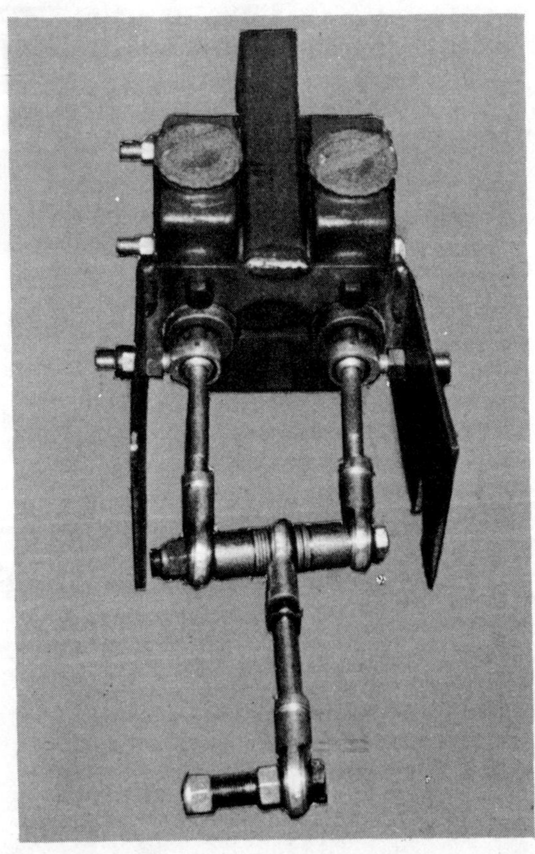

Mounting bracket for mounting two Edco or Girling master cylinders with balance bar. Push rod from brake pedal pushes on balance bar which right now is set [spaced by washers] to put more force on right hand master cylinder.

Left, the Hurst/Airheart stock car disc brake. Right, the H/A disc system coupled with a cutaway of a Norris Industries wheel specifically designed for disc brake application.

SPEEDWAY ENGINEERING DISC SYSTEM

Speedway Engineering calls their disc brake system the "Economy Disc Brake System" because it was designed to keep pace with the top quality stock car disc brakes, yet keep the prices in a budget range. The unit uses a 1.25-inch thick by 12.187-inch diameter cast iron disc, and employs a Kelsey Hayes four-piston non-floating caliper. The Kelsey Hayes caliper is one which was found on the 1965-67 Ford Galaxy, 1965-69 Lincoln, 1965-67 Thunderbird, and 1967-68 Eldorado and Toronado. It uses the D-1

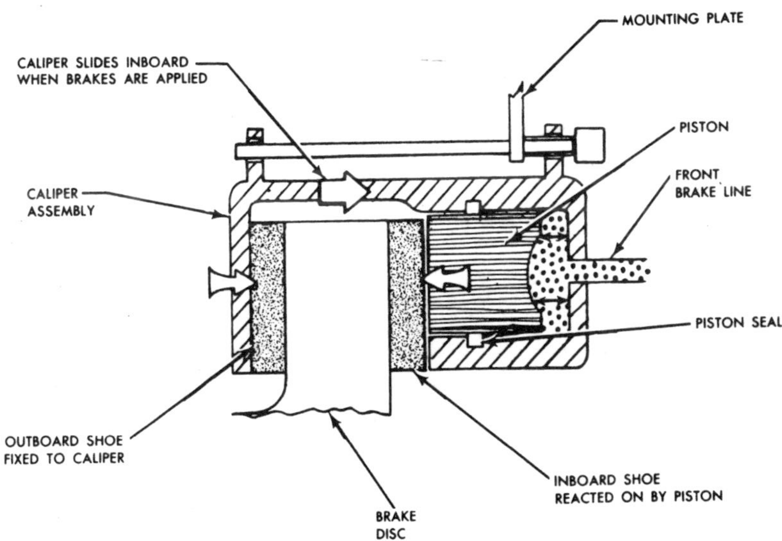

In the single piston caliper assembly, the hydraulic force is exerted in two directions: [1] against the piston surface, and [2] in the opposite direction against the caliper housing which pulls the housing inboard, sliding on rubber bushings. Drawing courtesy of Grey Rock.

This is the PRE single piston aluminum stock car caliper. Note the radial fins for strength and heat dissipation, and double square-shaped "O" rings. Fits mounts for GM, PRE, MRE and Howe brakes.

style of brake pad, which allows the racer to purchase racing linings from both Grey Rock and Bendix to fit it. The system uses the same rotors and calipers front and rear, to make standardization easy. To proportion for rear braking, a 1.25-inch wide strip of friction material is removed from the center of each pad in the rear.

HURST/AIRHEART

Hurst/Airheart, long time manufacturer of disc brake systems for street rods, drag cars and other lightweight competition machines, has introduced a product they call "The Stock Car Disc Brake." The calipers, rotors and adaptors are sold together as a complete package.

Hurst/Airheart worked long and hard to design a product which would run cool, not dissipate heat into the brake fluid (even with non-exotic fluids), and last through a 500-mile NASCAR Grand National race at Riverside or Martinsville. This product has proven that it meets the design challenge.

Part of its cooling secret, as we have mentioned previously in the book, is that it employs a massive rotor. It dissipates heat very well through increased ventilation area.

The calipers are of the non-floating design with four pistons per caliper, cast from aluminum. The rotors are cast iron. The brake lining material is Grey Rock's M-19, but made in a configuration which is unique to these brakes alone. This means pad replacements are available only through Hurst/Airheart and its authorized distributors. The lining material being used in this caliper is extremely thick, giving the pads a very long life expectancy.

PASSENGER CAR DISC APPLICATIONS

The two disc brake systems we have just discussed are very good systems, but like the old saying goes, "You get what you

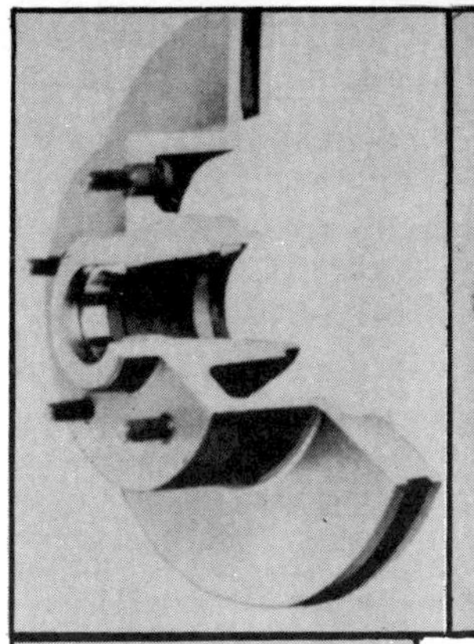

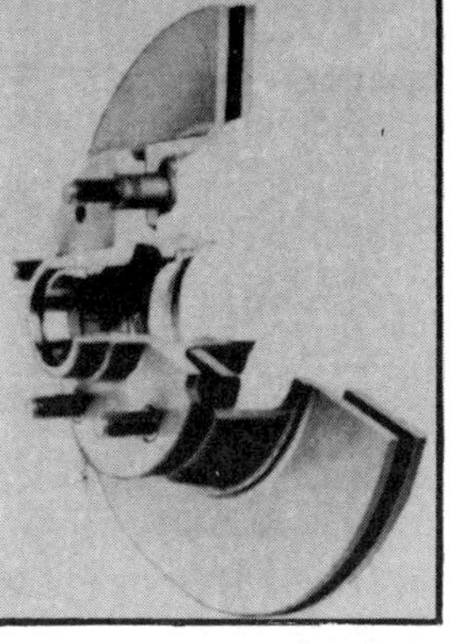

The three types of hub and rotor assemblies produced for passenger vehicles. At top left is the unicast or J-type, which has the hub and rotor as all one piece. At top right is the H-type which is a two-piece assembly but the hub is generally bolted or riveted to the rotor. At bottom left is the L-type, which has the hub and rotor easily dissassembled, being bolted together.

pay for." Well, the case with these two stock car racing disc brake systems is that they are exceptionally fine systems, which means the racer is going to be paying a pretty good chunk for either system. However, for short track racing applications, a system as exotic as these two is not required. In that case, something else is in order. The solution is the same as many others in a short track race car: adapt heavy duty components to racing application which can be readily found on passenger cars and light duty trucks. There are parts available from General Motors, Chrysler and Ford which can successfully be adapted to racing, and we will discuss the merits of each.

GENERAL MOTORS

Although the four-wheel Corvette disc brakes are widely known throughout the industry, they are not adequate for any race car heavier than a modified. And for that application, the brake assembly is too heavy at each wheel.

The best disc brake from General Motors to adapt to a racing vehicle is the 1971-1975 "Big Car Group" front discs. The application of this brake is the Chevrolet Impala and Caprice, Chevy

DELCO-MORAINE SINGLE PISTON DISC BRAKE

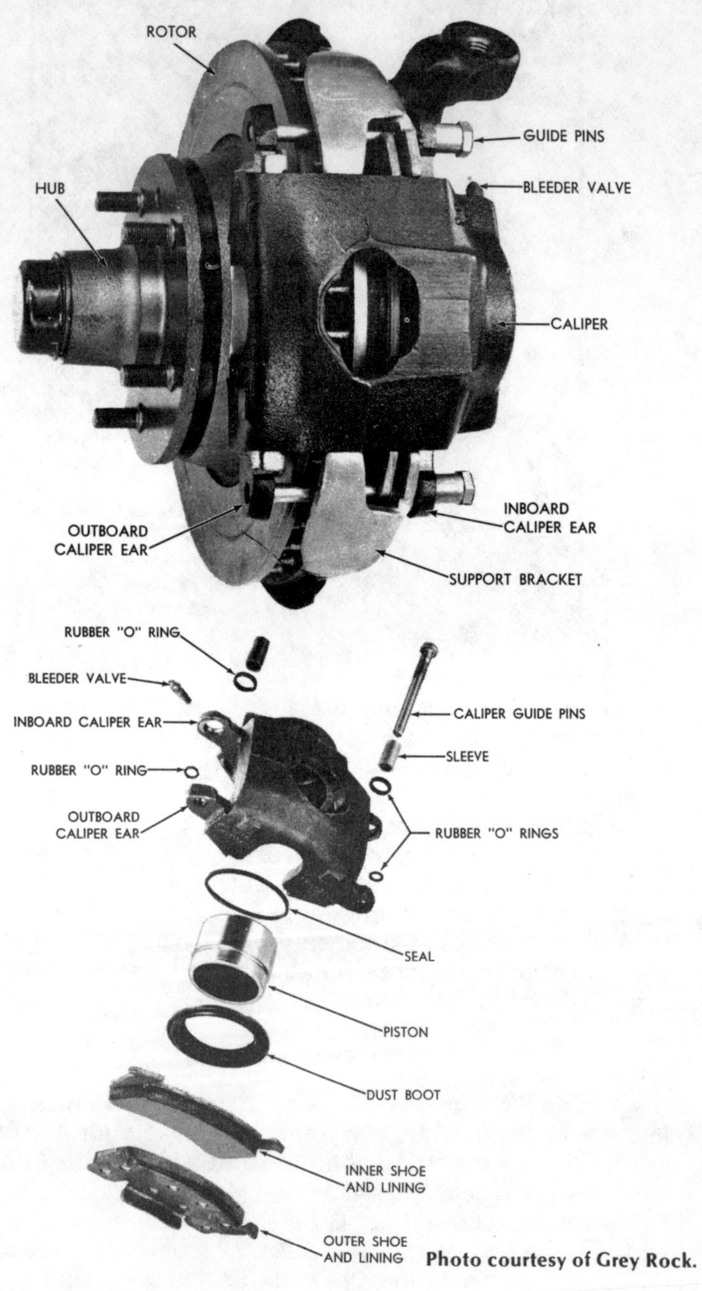

Photo courtesy of Grey Rock.

Blazer two wheel drive Series 10, half-ton truck (CP-1 and G-1), and 3/4-ton truck (G-20 and G-2).

The rotor for this disc brake is a unicast, or type J, rotor. This means the rotor and hub is cast together as one piece. The bolt pattern on it is 5 x 5 (which is a standard stock car racing wheel bolt pattern), and the wheel bearings which fit this hub are fairly decent in size.

If you care to use this disc brake system, but are opposed to the cast hub, there are two alternatives. First, the cast hub can be machined out of the assembly, holes drilled in the remaining flange on the rotor (remember to slightly slot the holes for heat expansion), and then a machined steel billet hub can be bolted to the rotor. A hub of this type (or any custom design from a steel billet) can be ordered from Speedway Engineering. This is the case with other cast hubs or J-type rotor and hub units we discuss later.

The second alternative is to use a different rotor with the existing caliper and hardware. This rotor, found in the 1973-75 Chevy Blazer Series 10 four wheel drive truck or 1971-75 Chevy K-1 ½-ton truck, is a two-piece hub and rotor assembly and can be purchased as a rotor only. The Chevy part number is 3995721, or the Raybestos Manhattan number is 5020. This rotor is 11.844" in diameter and 1.28" thick (same as the GM Big Car Group hub and rotor dimensions).

The caliper which fits these GM disc brake assemblies we have been discussing is Chevy part number 5474052 and -053. It is a cast iron single piston full floating caliper. It has a massive piston with 6.21 square inches of area.

Bendix Metal King brake pads are available to fit this caliper, as well as a Grey Rock M-19 material pad under part number HP-D52.

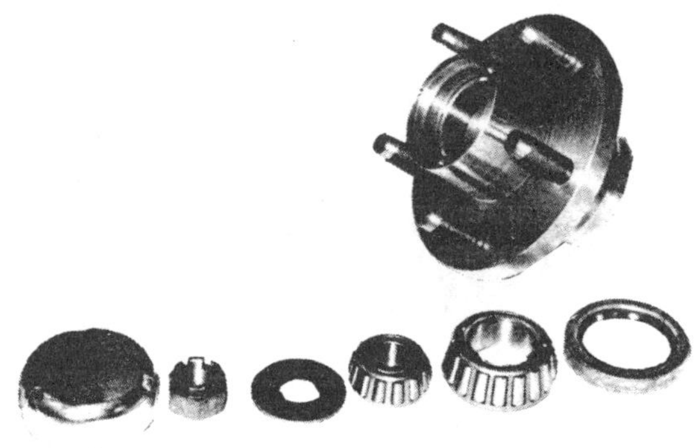

Speedway Engineering steel billet hub.

The best things to discuss in the Chrysler Corporation line-up are the four wheel disc brakes which Chrysler supplies for its "Kit Car" racers. The front brakes are from the 1973 Chrysler and other "C" body cars. The rear brakes are a 1971 Valiant front rotor and a 1969 Valiant caliper.

The front rotors Chrysler uses are part number P3690910. They are cast iron unicast J-type hub and rotor units. They are drilled for the common 5 x 5 bolt pattern. The 1973 Chrysler hub and rotor units are the same thing, available from Chrysler dealers under part number 3683969, BUT they are drilled for 5 x 4½ bolt pattern.

The front calipers are Chrysler part number 3744478 and -79. They are cast iron, single piston full-floating. Chrysler sells lining pads for these under part number P4007206. Bendix Metal King pads are also available for this caliper.

The rear rotor used on the Kit Car is a '71 Valiant because that particular model had a two-piece H-type hub and rotor which bolts

CHRYSLER SLIDING TYPE SINGLE PISTON DISC

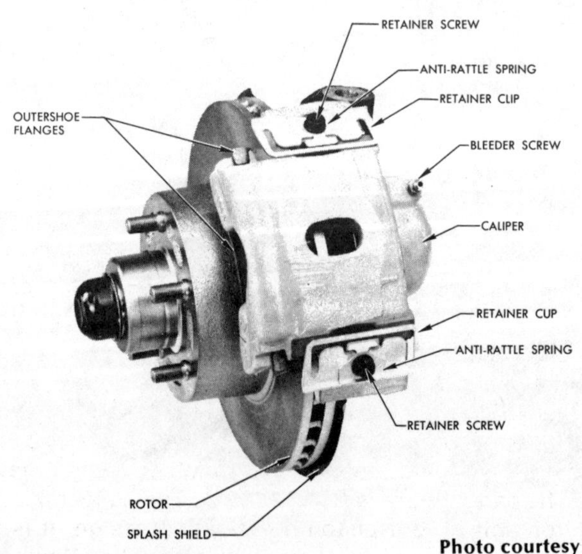

Photo courtesy of Grey Rock.

together. The rotor itself can be purchased under part number 3580502. Chrysler sells a hub which bolts to it for a Frankland quick change under part number P3690620.

The caliper used in the rear on the Kit Car is part number 2925220 and -21 which is a cast iron four piston non-floating unit.

There are a couple of alternative rotors available from Chrysler which make attractive units on which to build a disc system. In 1974 and 1975 the Imperial was equipped with disc brakes in the rear. The rotor, part number 3699699, is a K-type which bolts to a hub. The caliper, a cast iron single piston full-floating design, is part number 3699969 (without the pistons and seals). The rest of the unit includes adaptors number 3580778 and 79 and support number 3699731.

Dodge one-ton trucks from 1972 have been equipped with disc brakes in front, and the rotors are L-type which unbolt from the hub. The part number in 1972 is 3633118 and 19, and the part number for 1973 and later trucks is 3820186.

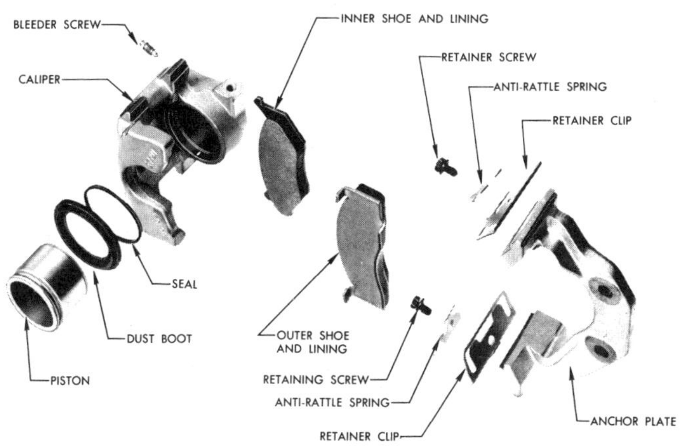

Photo courtesy of Grey Rock.

FORD MOTOR COMPANY

In 1969 Ford offered a special four wheel disc braking system for its Trans Am racing Mustangs, and these make the basis today for some good braking systems from Ford. The front brakes were from 1969 Lincoln and Thunderbird front disc assemblies, and the rear brakes were the front disc systems from the 1965-67 Mustang.

The 1969 Lincoln rotor is an H-type, which means it can be removed from the hub and a new hub of another design may be bolted to it. The part number is C9AZ-1102-A. The caliper is a very sturdy cast iron unit of four-piston non-floating design. It is part number C8AZ-2B120-A. Pads for it are available in Grey Rock M-19

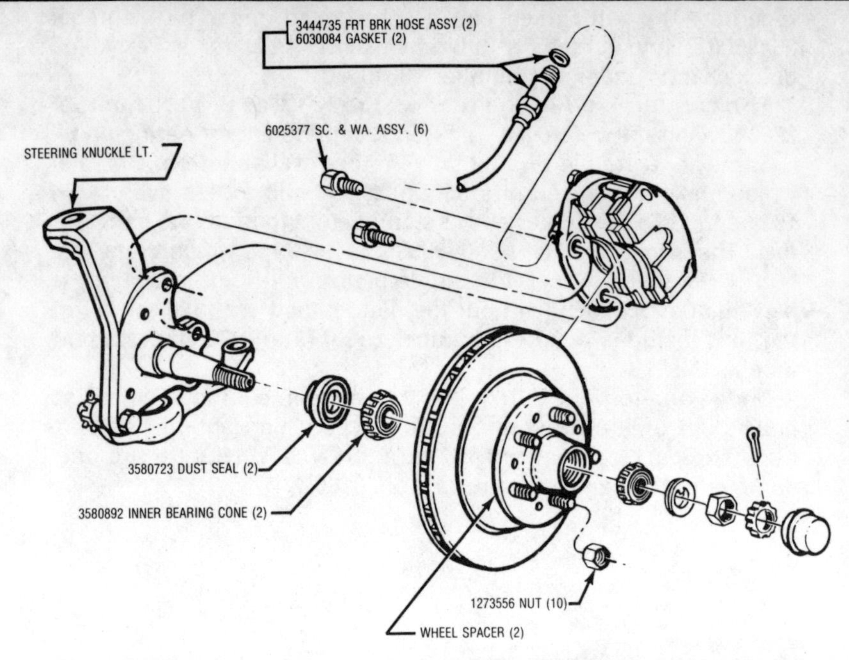

The Chrysler left front wheel disc brake assembly. Drawing courtesy of Chrysler Corporation Engineering Staff.

material under Grey Rock part number HP-D1, and from Bendix in the Metal King line.

The 1965-67 Mustang hub and rotor is also an H-type. It is part number C5ZZ-1102-B. The Mustang caliper is a four piston non-floating design, cast in cast iron. It accommodates Grey Rock M-19 material pads number HP-D11.

Another heavy duty rotor which is available from Ford is an L-type (rotor only) which fits the 1969 and later Ford F-350 one-ton truck front brakes. It is part number C8TZ-1102-D.

FITTING ROTORS

Many times it is possible to find a rotor in an H-type or L-type from one manufacturer and fit it to a caliper from another manufacturer. When doing this, be sure to check the new rotor overall diameter and thickness against the original application rotor dimensions. Generally, a caliper can accommodate a rotor which is as thin as .500" less than the original size and as small in diameter as .75" less than the original. If you mix and match rotors and calipers, though, be sure to carefully check the fit of all components before racing the vehicle.

In some H-type rotors, the hub is either bolted or riveted to the rotor. In some cases, some machining on the bolts or rivets is necessary before the rotor can be removed from the hub.

When adapting new hubs to rotors, or when attaching calipers to spindles with new adapting brackets, be sure to use only the best premium grade hardware as these parts are under extreme shear stress during braking.

This is the '69 Lincoln disc assembly which was adapted by Ford for its Trans Am racing Mustangs. Above are all the parts, and below installed on the front wheel.

KELSEY-HAYES FOUR PISTON DISC BRAKE

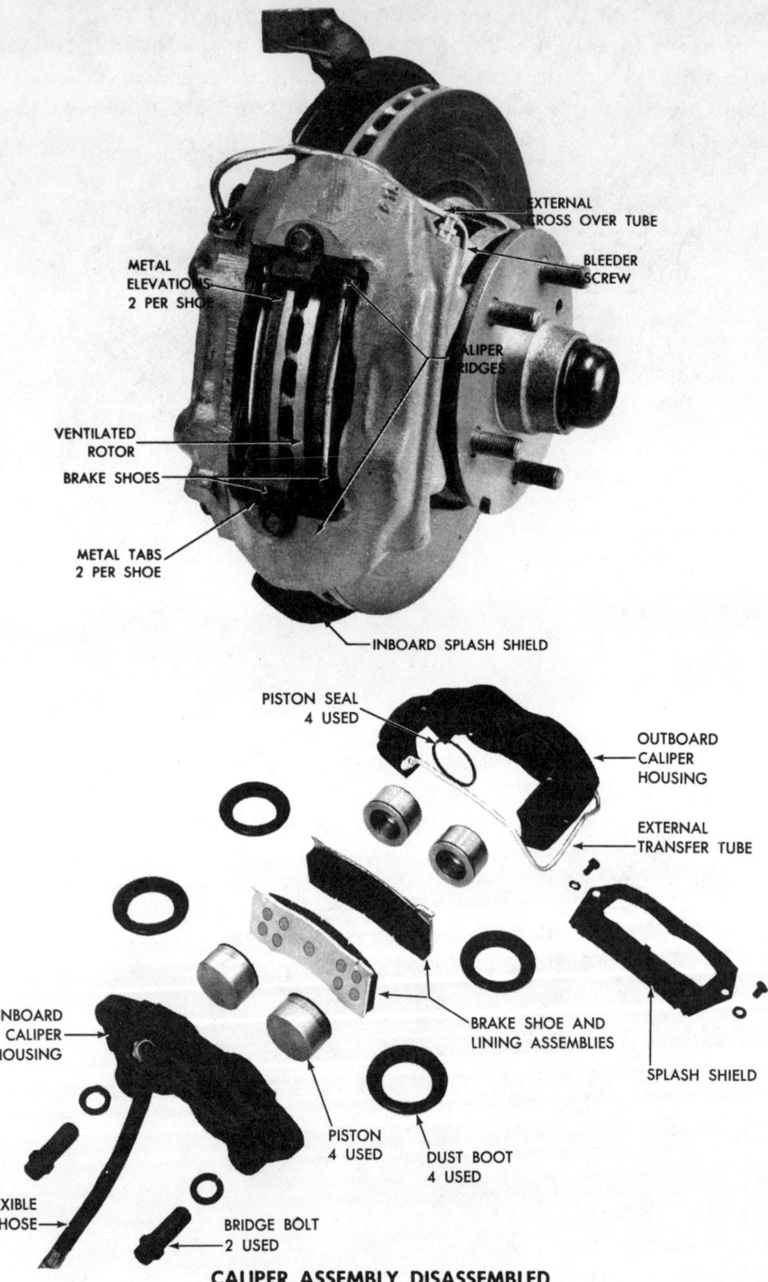

CALIPER ASSEMBLY DISASSEMBLED

Photo courtesy of Grey Rock.

PASSENGER CAR DISC SYSTEM SOURCES

ROTORS

Application	Mfr. Part #	RM #	Type	Bolt Pattern	Size
'71 Valiant	3580502	7202 (1)	H	5 x 4	.810 x 11.040
'73 Chrysler	3683969	7018	J	5 x 4½	1.250 x 11.750
'71-'73 Imperial	3580522	7014	H	5 x 5	1.250 x 11.750
'71-'75 Chevrolet	6274216	5006	J	5 x 5	1.250 x 11.860
'69 Lincoln Mark IV	C9AZ-1102-A	6008	H	5 x 4½	1.190 x 11.720
'65-'67 Mustang	C5ZZ-1102-B	6004	H	5 x 4½	.810 x 11.330
'73-'75 Blazer Series 10	3995721	5020	H	6 x 4½	1.250 x 11.860
'71-'73 Chevy Truck CR3	3995723	5014	L	(4)	N/A
'74-'75 Imperial (5)	3699699	7030	K	(4)	N/A
'72 Dodge 300 D Truck	3633118-19	7021	L	(4)	N/A
'73-'75 Dodge 300 D Truck	3820186	7023	L	(4)	N/A
'68-'75 Ford F-350 Truck	C8TZ-1102-D	6020	L	(4)	N/A

Notes:

(1) RM numbers are part numbers for replacement rotors available through Raybestos Manhattan and Grey Rock

(2) First number is width, second is diameter

(3) Not available.

(4) Part supplied as a rotor only without hub.

(5) Rear brakes.

CALIPERS

Application	Part #	Type	Lining
'69 Valiant	2925220-21	SPNF	3
'73 Chrysler	3744478-74	SPFF (2)	4, 5
'71-'75 Chevrolet	5474052-53	SPFF	5, 6
'68-'71 Lincoln	C8AZ-2B120-A	FPNF (1)	5, 7
'65-'67 Mustang	C5ZZ-2B120-C	FPNF	5, 8

Notes:

(1) Four piston non-floating
(2) Single piston full floating
(3) Chrysler part # P4007207
(4) Chrysler part # P4007206
(5) Bendix Metal King
(6) Grey Rock HP-D52
(7) Grey Rock HP-D1
(8) Grey Rock HP-D11

BRAKE BALANCE BARS

As we have discussed, it is important to design the brake system so that we can obtain maximum stopping power from all four wheels. As noted previously, under braking, weight is transferred to the front of the vehicle and therefore the front tires and the front brakes will be required to do more of the work. In a four wheel disc brake system we can get into the ballpark by choosing the proper piston diameters. This will allow us to achieve approximate front-to-rear brake balance. In order to fine-tune our ability to balance brake bias front to rear, it is necessary to use a dual master cylinder set-up with a balance bar unit. The balance bar system functions quite simply by altering the fulcrum point on a rod that connects the push rods to the front and rear master cylinders. By moving the fulcrum point we are able to change the amount of relative force that is applied to each master cylinder. This allows us to fine-tune the brake system so that the front brakes lock up just before the rear brakes.

Another advantage of using the dual master cylinder with balance bar set-up is one of improve safety by utilizing two master cylinders. Even if one system fails we still have partial braking force by utilizing the other system. A properly designed brake balance bar is the only way to obtain maximum efficiency from a disc brake system. Balance bar assemblies have become quite common in all forms of racing. Howver, many of the designs for brake balance bars are inadequate and can cause problems. Some of the things to be aware of in the design of a balance bar system are as follows:

1. The distance between the master cylinder push rods must remain constant and the push rods must maintain the same relationship to the master cylinder through full pedal travel.
2. The master cylinders must be able to complete their full stroke without the balance bar binding on the side of the pedal.
3. The balance bar shaft must remain in a plane horizontal to the position of the master cylinders at all times during travel. Because of this, spherical rod ends are not satisfactory for the push rod pivots. The rod ends will pivot vertically and bind the pedals and restrict the travel of the push rods to the master cylinder.

Excellent brake balance bar assemblies are available from Tilton Engineering and plans for a brake balance bar are included in the disc brake blueprints available from Steve Smith Autosports. The only way to achieve the proper brake balance in a race car is to utilize dual master cylinders with a brake balance assembly, or to use a brake proportioning valve. In some instances (primarily applications on dirt track cars) the use of a brake balance bar with dual master cylinders will allow front to rear proportioning and the use of the brake proportioning valve will allow the driver to adjust left side to right side brake balance for changing track conditions. The proportioning valve should be plumbed between the front brakes on dirt track cars.

SOME NOTES ON ADJUSTING BRAKE BALANCE

1. Changing the wedge or ride height of a car changes braking characteristics and generally will cause brake balance to be altered.
2. Brake balance bars cannot be adjusted in the shop by turning the wheels and engaging the brake system. The only way to determine brake proportioning in the shop is to use pressure gauges with a minimum rating of 2,000 pounds, attached to the front and rear brake lines.
3. The balance bar push rod should be adjusted so that the balance bar is perpendicular to the master cylinder push rods when the pedal is released. When the pedal is released, both push rods should be pulling against the retaining washers and the master cylinders. If they are not, the push rods can then advance the piston enough to cover over the bleed hole in the caliper and this will cause the brakes to lock up as soon as they get hot.
4. The only way to balance a brake balance bar assembly is by changing the fulcrum point between the master cylinder push rods and the pedal. In no master cylinder set-up with the balance bar can the proportioning be adjusted by changing the length of the push rods relative to each other.

WHEEL CLEARANCE

In mounting some disc brake systems the problem of caliper to wheel clearance will be encountered. There are two alternatives: first, use a wheel with greater offset. Second, purchase a wheel which is made especially for use with disc brakes.

Friction Materials

We will discuss the leading friction materials available from the major brake materials companies, then we will get specific with applications of these materials. While we discuss friction materials (or linings or pads, or whatever you wish to call them), remember that in a racing situation, they must possess four characteristics: (1) a retention of excellent coefficient of friction, (2) ability to withstand extremely high temperatures, (3) a smooth and stable operating condition to prevent periodic pulling, and (4) extreme resistance to wear.

When choosing lining materials, remember that the harder the linings, the higher the resistance of the lining to fade, BUT also, the harder the linings, the more force required on them. This means a harder lining needs a smaller master cylinder.

GREY ROCK MATERIALS

Grey Rock manufactures two drum brake friction materials and one disc brake friction material for racing applications. Grey Rock 5262, one of the two drum brake materials, is a full metallic lining (meaning it does not contain any organic material such as asbestos). The greatest characteristic of 5262 is that it holds extremely constant with no wheel lock-up or pull. 5262 was first developed as a superspeedway lining for Grand National cars, where wheel lock-up could be extremely dangerous.

The second drum material by Grey Rock is known as 5191. It is also a full metallic lining, but it holds better than 5262 and withstands heat much better than 5262. 5191 was developed for Grand National cars running tracks like Martinsville and Riverside where brake wear and fade is rapid and common. Grey Rock cautions about using 5191 on anything but a short track application because the lining has some swelling to it as it gets hot, and this coupled with a bent anchor pin could cause wheel lock-up.

Grey Rock 5262 lining which has just been broken in during practice. Notice the stainless steel braided flex line.

Grey Rock's disc brake high performance linings are made of a material commonly called M-19, but designated as 4528-19-FF on the box. The material is semi-metallic, but possesses the same characteristics as Grey Rock 5191. The M-19 has shown itself to wear very well and withstand extremely high operating temperatures without loss of stopping efficiency.

BENDIX MATERIALS

Bendix makes a full organic drum brake lining known as EDF, but it has shortcomings in any application where the brakes are used a great deal. It would probably work with very lightweight stock cars and hobby class cars. Bendix makes EDF linings for use in taxis and police cars where high temperatures and wear are a problem.

Bendix is making a semi-metallic disc brake lining which is very suitable for racing application. Called "Metal King," it depends on a high heat level to operate properly. A high heat level with discs could be defined as 1000 to 1200 degrees F at the rotor face, or 700 to 900 degrees F at the lining face. These are temperatures encountered by racing disc systems in the most severe cases, such as lap after lap of hard driving in a Grand National car at Riverside.

Metal King linings are very versatile for the racer who adapts a passenger car disc system to his race car because the linings are readily available in sizes to fit most all passenger car calipers.

VELVETOUCH MATERIAL

Velvetouch is a product name for a semi-metallic drum brake lining manufactured by the S.K. Wellman Corp. Velvetouch is a popular choice of short track racers, mostly because it is one of the most readily available drum materials throughout the country. It stops even the heaviest Grand National cars well, but its major drawback is that it is a bonded rather than riveted lining. The problem is that the maximum operating temperature of the lining is greater than that which the bonding agent can endure, so many racers using Velvetouch linings find that the linings will come off the shoes before they completely wear out. Many people have tried riveting Velvetouch linings to the shoes, but the material itself does not have enough body to it. Friction force from the drums will pull the material loose from the rivets.

DELCO MORAINE LINING

There is a semi-metallic lining with characteristics very similar to the Velvetouch lining which is available from Chevrolet dealers under part number 3830635. The lining is very good for short track racing, having very even wear and friction qualities.

INSTALLING DRUM LININGS

To take full advantage of any racing friction material, you must be sure to follow proper preparation techniques of the linings and drums, or else optimum stopping will not be achieved.

First, make sure that drums are turned at a .004" feed on the brake lathe for a perfect drum finish. After the drums are machined, wash them with a laundry detergent and rinse thoroughly. It is best to use compressed air to blow them dry. The washing removes tiny metal chips and small particles of graphite from the drum. After the drums have been washed, handle them with care.

The Delco Moraine semi-metallic lining.

Grease or body oils on the newly machined surface can cause erratic braking action after the drums and brakes have been assembled.

Next, the linings should be ground to match the drums. Some brake lining grinders automatically calibrate the amount to grind. If the one you are using does not, remember that the brake linings should be cut .030" under the drum diameter. So, if the drum mikes .040" over the original size, for example 11.040", then the grinder should be set for 11" plus .010."

The lining should have .008" to .010" clearance at the toe and heel of the lining. This will give a slight rocking of the toe and heel to the drum. Be sure to check the fit of each shoe in the drum. Never let the toe and heel contact the drum with clearance in the center of the lining. The pedal will feel firm, but the car will never stop.

After the brakes are assembled, it is important to properly "seat in" the linings before the car is raced. This process will require about 10 laps around most short tracks, with the driver applying the brakes moderately about four times each lap. After this is done, return the car to the pits and allow the brakes to cool. Then repeat this process and again cool the brakes. The linings should by this time be properly seated, and there will be a difference in the feel of the pedal. If the linings are not broken-in correctly, they can cause fade the first time they generate extreme heat.

DISC BRAKES PADS

Most manufacturers of friction materials for disc brake pads make pad material in three distinct compounds. These fall into the general categories of soft, medium and hard linings. In choosing the proper material for disc pads, consult with the manufacturer of your system. A representative of the manufacturer should be able to help you to determine the proper pad compound for your particular application. Using the Hardie-Ferodo line of disc pad material as an example, the relative differences between the soft, medium and hard linings and their applications will become apparent. Hardie-Ferodo's soft lining is known as premium. This compound is suitable for cars that have difficult in getting a harder racing friction material up to temperaturee. It will generate maximum stopping power but will have a very short pad life. This compound is generally used on lightweight cars or on cars that race on long tracks where little braking is used and long periods of time pass between applications of the brakes. This lining develops maximum friction immediately upon application of the brakes without causing a high rate of rotor wear.

The medium compound in Hardie-Ferodo's line is known as DP11. This is a metallic compound with a high copper content and is very similar to the Ferodo DS11 or the Raybestos 19-M in performance. This material is designed for high fade resistance and will still offer reasonable pad wear without excessive pedal pressure. This is the most common type of brake friction material and is used over a wide range of racing applications. The friction coefficient is stable over a wide change of brake pad temperatures and promotes uniform braking application. With this lining, brakes will remain fade-free until the rotor overheats. Rotor life is generally exceptional with this type of lining. The hard compound in Hardie-Ferodo's line is known as 1103 Compound. This compound was primarily developed for endurance racing for use in heavy sedans at places like Sebring, Le Mans and Daytona. This compound will deliver between 30 and 50% more pad life than the DP11 without excessive rotor wear. Although pad life is increased, so is the pedal pressure. In long distance races where pad life is critical this type of compound has proven very effective.

The use of different hardness pads to alter the front to rear brake bias should be avoided. Pads of different hardness have different braking characteristics between hot and cold temperatures. A hard pad usually takes longer to reach maximum effectiveness because it is designed to operate at high temperatures. A soft pad will be effective immediately. Since brake pads will heat up at different rates, this can alter the effectiveness of the braking system tremendously at different rates depending upon the temperature of the system. This tends to create an imbalance and causes the driver to have to compensate for the brake system which is less than an ideal situation.

INSTALLING DISC PADS

When tearing down a disc brake system for pad replacement, check the pads for tapered or uneven wear. This wear can indicate caliper flex or spindle flex. Uneven brake pad wear may also be caused by poor bracket design or flex of the brackets. If the pad selected is either too hard or too soft for a given application, taper of the pad can be caused not necessarily by wear but by the transfer of friction materials from one end of the pad to the other. This causes the pad to be thicker at one end and thinner at the other.

Check the rotor. A severely scored rotor can impair the operating efficiency of the braking system. A scored rotor should ideally be replaced. Measure the thickness of the rotor with a micrometer at six evenly spaced intervals on the friction surface area. The thickness should not vary any more than .0005-inch (Yes, that very minute figure is correct.). If it does vary more than this, replace the rotor. Any larger variation can cause high speed brake chatter and loss of braking efficiency. It is also a good idea to check the rotor for run out and taper at the same time. When installing new disc pads it is important to utilize proper break-in procedure.

New pads experience a phenomenon known as green fade. This occurs when new brake pads are heated to their operating temperature and the bonding agent that holds the friction materials together becomes gaseous and causes a gas to fill the area between the friction material and the rotor. This reduces the coefficient of friction tremendously and causes the phenomenon known as green fade. In order to eliminate this it is important to properly break-in disc pads.

Pads are broken in by alternately heating and cooling them while running on the track. The brakes should be used relatively hard in a few applications and then allowed to cool down completely. This should be done two or three times to insure that the brakes have been properly bedded or broken in. A visual inspection of the brake pads will determine if they have been broken in properly. The outer surface of the pad should have a grayish color, different from the rest of the pad. THis should extend down into the pad between 1/8 and 1/4-inch from the braking surface. It is important to keep in mind that you should never start a race with new pads, unless they are properly broken in. The phenomenon of green fade will continue until the brakes have had an opportunity to cool down.

On the following pages are illustrated the most common disc brake pad shapes. Then they are listed by make and model application, then are converted from the common FMSI numbers to Bendix Metal King pad part numbers.

FMSI numbers are standardized reference numbers for brake linings. They are the property of and are copyrighted by Friction Materials Standards Institute Inc., Paramus, New Jersey 07652.

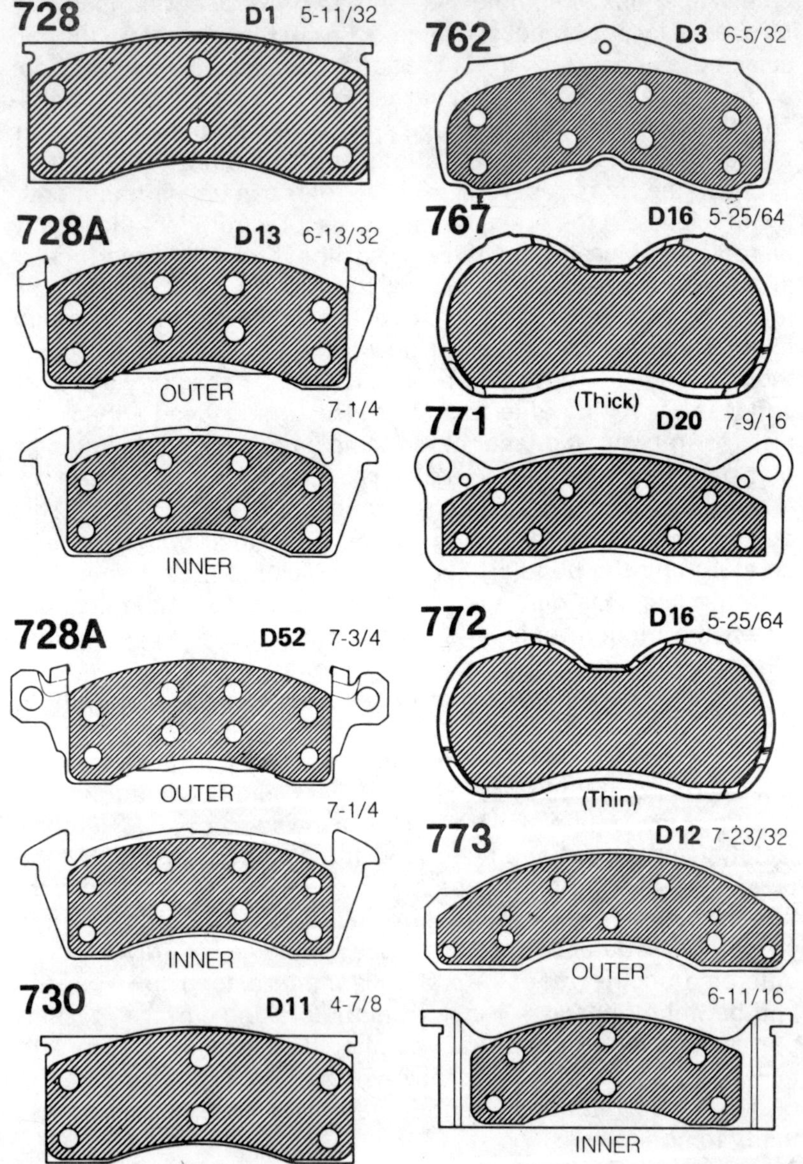

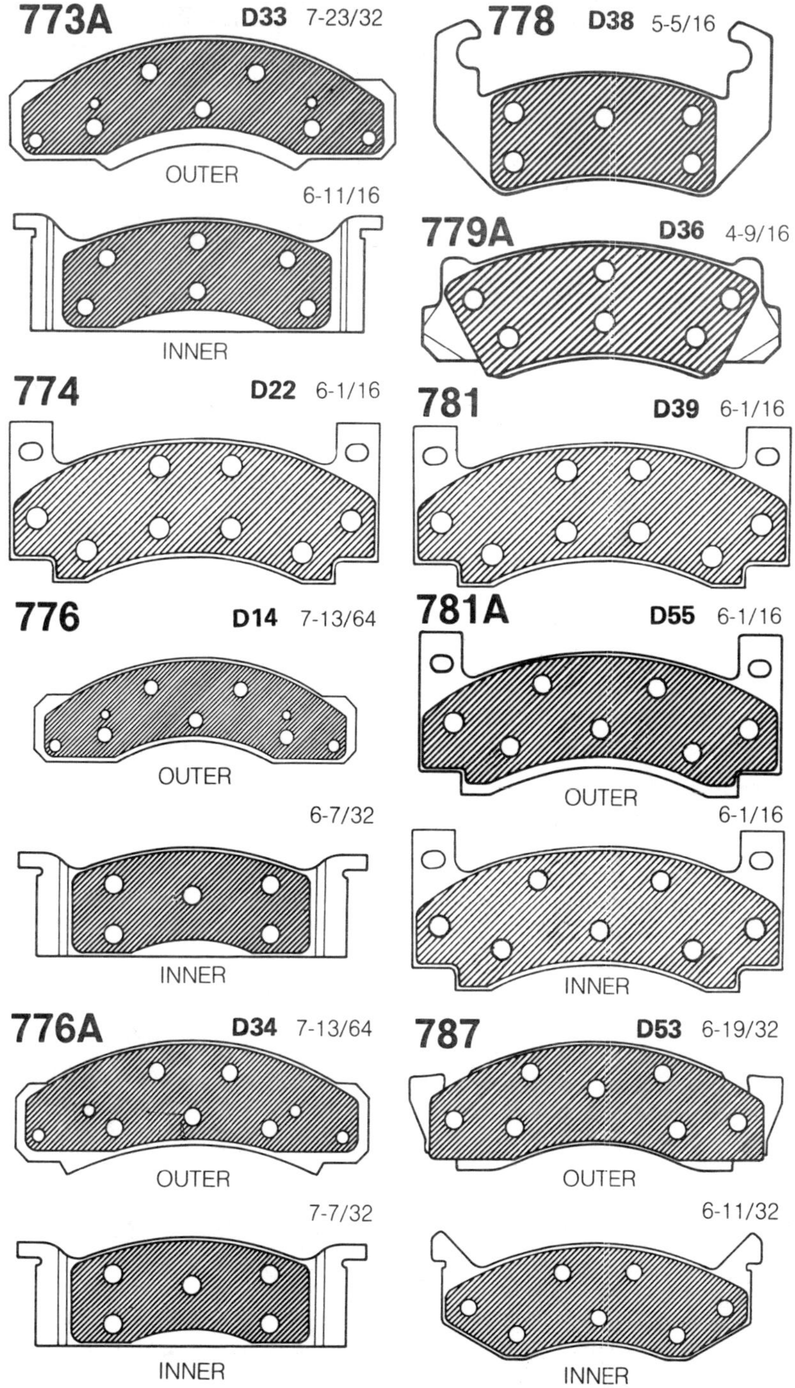
773A
D33 7-23/32
OUTER
6-11/16
INNER
778
D38 5-5/16
779A
D36 4-9/16
774
D22 6-1/16
781
D39 6-1/16
776
D14 7-13/64
OUTER
6-7/32
INNER
781A
D55 6-1/16
OUTER
6-1/16
INNER
776A
D34 7-13/64
OUTER
7-7/32
INNER
787
D53 6-19/32
OUTER
6-11/32
INNER

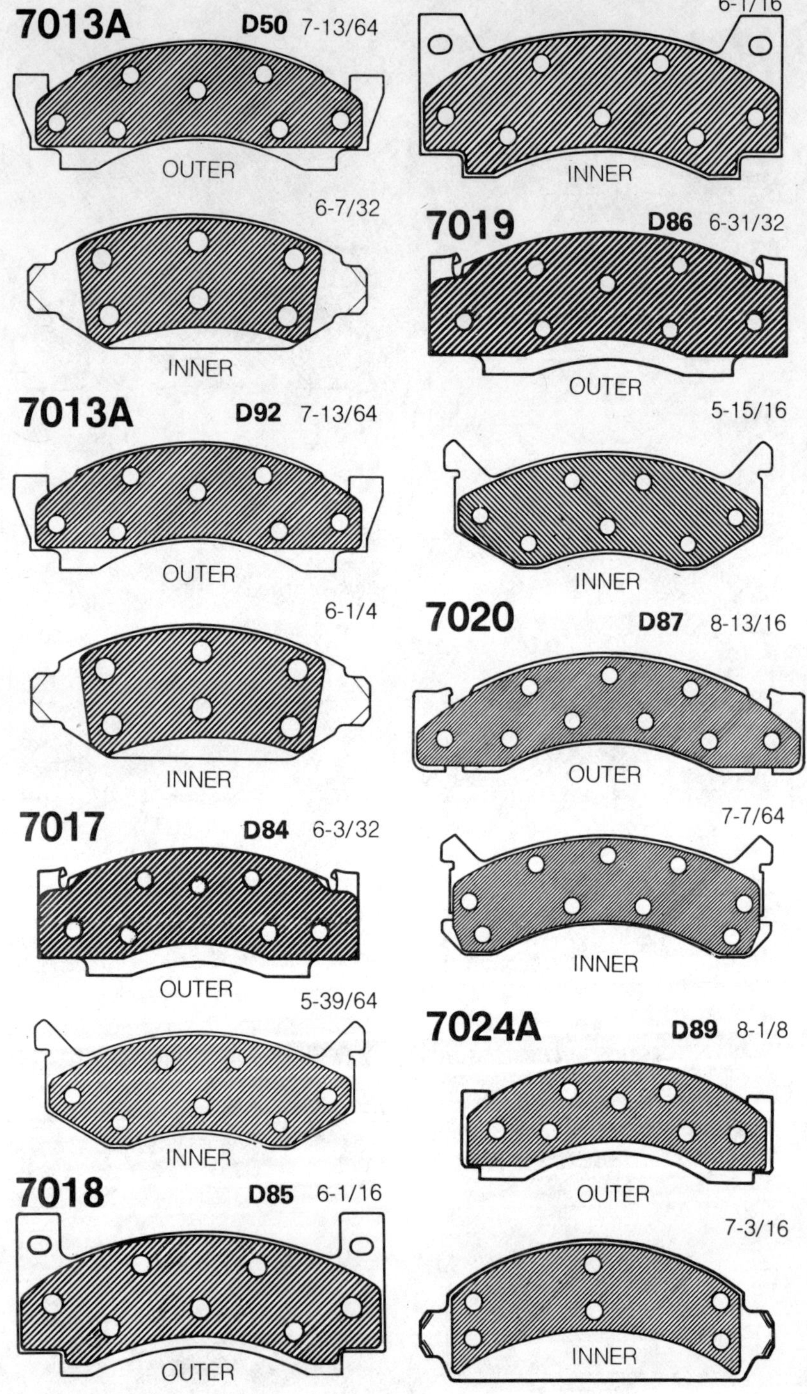
7013A
D50 7-13/64
OUTER
6-7/32
INNER
7013A
D92 7-13/64
OUTER
6-1/4
INNER
7017
D84 6-3/32
OUTER
5-39/64
INNER
7018
D85 6-1/16
OUTER
6-1/16
INNER
7019
D86 6-31/32
OUTER
5-15/16
INNER
7020
D87 8-13/16
OUTER
7-7/64
INNER
7024A
D89 8-1/8
OUTER
7-3/16
INNER

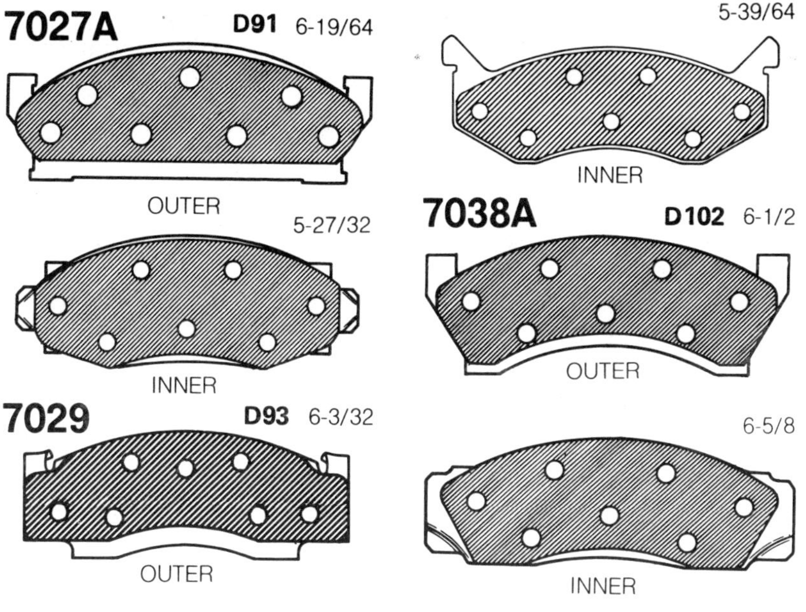

Year	Make and Model	FMSI No.
	AMERICAN MOTORS	
1975	All	7013A D-92
1974-72	All	781A D-55
1971	All	781 D-39
1970-65	All	731 D-18
	BUICK	
1975	Skyhawk	778 D-38
1975-72	All	728A D-52
1971-69	All	728A D-13
1969-68	Buick	772 D-16
1968-67	Special	762 D-3
	CADILLAC	
1975-72	All	728A D-52
1971-69	All	728A D-13
1968-67	Eldorado	728 D-1
1968-67	Cadillac	728A D-13
	CHECKER MOTORS	
1975-73	All	728A D-52
1972-70	All	728A D-13
1969	All	762 D-3
	CHEVROLET	
1975-72	All	728A D-52
1971-69	All	728A D-13
1968-67	All	762 D-13
	CAMARO, CHEVELLE, CHEVY II/NOVA, MONTE CARLO & EL CAMINO	
1975-72	All	728A D-52
1971-70	El Camino	728A D-13
1971-69	All	728A D-13
1970-69	Camaro w/4whl disc	727 D-8
1968-67	All	762 D-3
	CORVETTE	
1975-74	All F & R	7028A D-8
1973-65	All F & R	727 D-8

Year	Make and Model	FMSI No.
	VEGA	
1975-71	All	778 D-38
	CHEVROLET TRUCK	
1975-74	C-, G-, P-30	7024A D-89
1975-74	C-, G-, K-, P-10; C-, G-, K-, P-20; C-, G-, P-30	728A D-52
1975-73	All (½-T)	728A D-52
1972-71	G-10, G-20, P-20; C-, P-30	728A D-13
	CHRYSLER, IMPERIAL & CORDOBA	
1975	Cordoba	7018 D-85
1975-74	Imperial (Rear)	7029 D-93
1975-74	Chrysler & Imperial (F)	7019 D-86
1973	All	7018 D-85
1972-71	All	781 D-39
1970	Imperial	732 D-22
1970-69	Chrysler	774 D-22
1969-66	Imperial	732 D-10
1968-66	Chrysler	732 D-10
	DODGE, CHALLENGER & CHARGER	
1975-74	Dodge	7019 D-86
1975-73	Challenger & Charger	7018 D-85
1973	Dodge	7018 D-85
1972	All	781 D-39
1971-70	All	774 D-22
1969-66	All	751 D-19
1968-67	Dodge	732 D-10
	DART	
1975-73	All	7017 D-84
1972-65	All	730 D-11
1969	GTX & Tow Pak	751 D-19
	DODGE TRUCK	
1975	500 Series	7020 D-87
1975	W-200, 300	7024 D-89
1975-73	AD,AW,B-D-100;B,D-200;B,D,R,BF,M-300	7019 D-86
1972	B-D-100,B-D-300	781 D-89
1972	D-200, D-300	787 D-53
1971	B-100, B-200	774 D-32

	FORD	
1975-73	All	7013A D-50
1972-71	All	773A D-33
1970-69	All	773 D-12
1968	All	773A D-33
1967-65	All	728 D-1
	GRANADA & MAVERICK	
1975	All	7027A D-91
	FAIRLANE/TORINO, FALCON & RANCHERO	
1975-72	All	7013A D-50
1971	All	776A D-34
1970-69	All	776 D-14
1968	All	776A D-34
1967	All	730 D-11
	MUSTANG & MUSTANG II	
1975-74	All	7026A D-90
1973-71	All	776A D-34
1970-69	All	776 D-14
1968	All	776A D-34
1967-65	All	730 D-11
	PINTO	
1975-74	All	7026A D-90
1973-71	All	779A D-36

Year	Make and Model	FMSI No.
	OLDSMOBILE	
1975	Starfire	778 D-38
1975-72	All	728A D-52
1971-69	All	728A D-13
1968-67	All Exc. Toronado	762 D-3
1968-67	Toronado	728 D-1
	PLYMOUTH	
1975	Grand Fury	7019 D-86
1975-73	Fury & Satellite	7018 D-85
1974	Fury	7019 D-86
1972	Fury & Satellite	774 D-22
1971-70	Belvedere & Satellite	774 D-22
1971-69	Fury	774 D-22
1969-66	All Exc. Fury	751 D-19
1968-66	Fury	732 D-10
	BARRACUDA, DUSTER & VALIANT	
1975-73	Duster & Valiant	7017 D-84
1974-73	Barracuda	7018 D-85
1972-71	Barracuda	781 D-39
1972-70	Duster & Valiant	730 D-11
1970	Barracuda	774 D-22
1969-65	All	730 D-11
	PLYMOUTH TRUCK	
1975-74	All	7019 D-86
	PONTIAC	
1975	Astre	778 D-38
1975-72	All	728A D-52
1971	Firebird	728A D-52
1971-69	All Exc. Firebird	728A D-13
1968-67	All	762 D-3

	THUNDERBIRD	
1975	All (Rear)	7038A D-102
1975-72	All	7013A D-50
1971	All	773A D-33
1970-69	All	773 D-12
1968	All	773A D-33
1967-65	All	728 D-1
	FORD TRUCK	
1975-72	F-100, F-250 6,200 GVW	7013A D-50
1975-68	F-250 8,100 GVW	771 D-20
	GMC TRUCK	
1975-74	3500 9,000-10,000 GVW	7024A D-89
1975-72	Sprint, 1500, 2500, 3500	728A D-52
1971	1500,2500,3500 C,G,P,K	728A D-13
	IHC TRUCK	
1975-74	Scout II, 100,200 Series	7013A D-92
1975	200 HD, 1300 Series	7024A D-89
	JEEP	
1975-74	Wagoneer & Cherokee, Models 25,26,45,46	728A D-52
	LINCOLN	
1975	All (Rear)	7038A D-102
1975-73	All	7013A D-50
1972-71	Lincoln, Mark III	773A D-33
1972	Mark IV	7013A D-50
1970-69	Mark III	773 D-12
1969-65	All Exc. Mark III	728 D-1
1968	Mark III	773A D-33

	MERCURY & MONARCH	
1975	Monarch	7027A D-91
1975	All (Rear)	7038A D-102
1975-73	All (Front)	7013A D-50
1972-71	All	773A D-33
1970-69	All	773 D-12
1968	All	773A D-33
1967-66	All	728 D-1
	COMET, COUGAR, MONTEGO	
1975-74	Cougar	7013A D-50
1975-74	Comet	7027A D-91
1975-72	Montego	7013A D-50
1973-71	Cougar	776A D-34
1971	Montego & Cougar	776A D-34
1970-69	Montego & Cougar	776 D-14
1968	All	776A D-34

GREY ROCK CROSS REFERENCE

FMSI NO.	GREY ROCK NO.
728 D-1	HP-D1
728A D-13	HP-D52
728A D-52	HP-D52
730 D-11	HP-D11
732 D-22	HP-D39
774 D-22	HP-D39

DISC PAD INTERCHANGE

Specified Set	Can Also Use	Specified Set	Can Also Use	Specified Set	Can Also Use
728A D-13	728A D-52	**773** D-12	773A D-33	**776** D-14	776A D-34
768 D-12	773A D-33	**774** D-22	781 D-39	**778** D-35	778 D-38
769 D-14	776A D-34			**788** D-50	7013A D-50

BENDIX CROSS REFERENCE

FMSI NO.	BENDIX NO.	METAL KING	FMSI NO.	BENDIX NO.	METAL KING
727 D-8	318175	—	778 D-38	325003	448061
728 D-1	321659	—	779A D-36	325000	—
728A D-13	322976	448041	781 D-39	323789	448045
728A D-52	322976	448041	781A D-55	323789	448045
730 D-11	317514	—	787 D-53	—	—
731 D-18	325004	—	7013A D-50	448006	448044
732 D-10	321173	—	7013A D-92	448048	—
751 D-19	318636	—	7017 D-84	448038	—
762 D-3	321389	—	7018 D-85	323789	448045
767 D-16	322785	—	7019 D-86	448022	448055
771 D-20	323109	448059	7020 D-87	—	—
772 D-16	322785	—	7024 D-89	448039	448062
773 D-12	322596	448042	7026A D-90	448047	—
773A D-33	322596	448042	7027A D-91	448046	—
774 D-22	323789	448045	7028A D-8	318175	—
776 D-14	325009	448043	7029 D-93	448038	—
776A D-34	325009	448043	7038A D-102	448060	—

BRAKE PARTS AND HARDWARE SUPPLIERS

AP Racing
Lemington Spa
Warwickshire, England
44926-27000

Bendix Automotive Corp.
1904 Bendix Drive
Jackson, TN 38301
(901) 423-1300

Delco Moraine Div.
General Motors Corp.
Dayton, OH

Down Corning Corp.
Midland, MI 48640
(517) 636-8000

Earl's Supply
14611 Hawthorne Blvd.
Lawndale, CA 90260
(213) 679-1438

Edco Specialty Products
3411 W. MacArthur Blvd.
Santa Ana, CA 92703
(714) 540-6536

EIS Automotive Corp.
West Bend, WI

ELSCO Inc.
1843 E. Adams St.
Jacksonville, FL
(Ferodo Pads)

Girling Braking Systems
Div. Lucas Electrical
200 Manchester Ave.
Detroit, MI

Grey Rock Div.
Raybestos Manhattan Corp.
P.O. Box 9140
Bridgeport, CT 06603
(203) 371-0101

Hurst Airheart
40 West Street Road
Warminister, PA 18974
(215) 672-5000

JFZ Engineered Products
7851 Alabama Ave., Unit 3
Canoga Park, CA 91304
(213) 887-6776

Midwest Race Engineering **(MRE)**
503 S. 7th St.
Boonville, IN 47601
(812) 897-0970

Plasma Technology Inc.
1754 Crenshaw Blvd.
Torrance, CA 90501
(213) 320-3373

Professional Racers Emporium **(PRE)**
1463 E. 223rd St.
Carson, CA 90745
(213) 830-4678

Speedway Engineering
13040 Bradley Ave.
Sylmar, CA 91342
(213) 362-5865

Tilton Engineering
P.O. Box 1787
Buellton, CA 93427
(805) 688-2353

Troutman, LTD.
3198 'L' Airport Loop Dr.
Costa Mesa, CA 92626
(714) 979-3295

Veletouch Div.
S.K. Wellman Corp.
200 Egbert Rd.
Bedford,, OH 44146
(216) 232-2400

Wilwood Racing Brakes
9666 Owensmouth Ave., Unit S
Chatsworth, CA 91311
(213) 346-4126

Winters Performance
2819 Carlisle Rd.
York, PA 17404
(717) 764-9844

Yankee Silicones
P.O. Box 1089
Schenectady, NY 12301
(518) 370-4177

Other books from Steve Smith Autosports

To order any book listed, specify book number and send full price to:
Steve Smith Autosports, P.O. Box 11631, Santa Ana, CA 92711

The Buick Free Spirit Power Manual

Complete information on the performance engine of the future – the Buick V6. Blocks, crankshafts, rods, valve gear, intake and exhaust systems, ignition, suspension, brakes and body modifications. Complete list of performance hardware. Over 200 photos and drawings. **Order #S123...$6.95**

Stock Car Driving Techniques

Written with detail and authority by 1975 Daytona 500 winner Benny Parsons. Chapters include: Basic competition techniques • Getting started in racing • Developing a proper driving style • Driving the dirt • Improving your lap times • Defensive driving techniques • Superspeedway driving details. Covers every aspect of competition driving. 96 pages. **Order #S104...$7.45**

The Stock Car Racing Chassis

This is a basic handling and suspension book with facts on caster, camber, toe-out, tire temperatures and sorting a chassis. Special feature is the well-illustrated chapter on building a competitive Chevelle chassis. **Order #S101... $6.00**

The Complete Stock Car Chassis Guide

A handling and suspension book discussing racing tire principles and selection, dirt track chassis design, roll cage construction, shock absorber selection, how to build the two-point suspension, aerodynamics, roll couple, anti-roll bars and much more. Packed with photos and drawings. **Order #S102... $6.00**

The Best Of Chevy Hot Line

This book is a compilation of articles from 3 years' worth of Chevy Hot Line issues. The articles have been excerpted and grouped under the general topic discussed, such as cams, pistons, carburetion, etc. You'll find this to be a real engine building and trouble-shooting reference work. **Order #S124... $10.50**

Building the Hobby Stock/Street Stock Car

This book is written for the low-buck racer with limited facilities and who wants to farm out a minimum of work. It shows you how to build a car with a minimum of "store-bought" parts or expensive machine work. Chapters include: your choice of car • cage construction • chassis vs. unitized car • engine, cooling and electrics • transmissions and rear ends • tires and wheels • suspension (parts choice, building and sorting) • driving • spare parts and preparation. **Order #S126... $6.45**

Advanced Race Car Suspension Development

The latest technological information about race car suspension design and development. A former GM chassis engineer reviewed it: "This is the most complete and accurate suspension book ever published." The book details everything about race car suspension design, dynamic chassis analysis, chassis structure, handling stability and testing techniques. 176 pages, over 100 photos and drawings. **Order #S105... $6.95**

Chevy Heavy Duty Parts List

It is up-to-date with all of the newest stuff from the factory. Complete with all engine and driveline goodies, plus much more miscellaneous info for the racer. **Order #S120...$3.50**

SSA Chassis & Cage Blueprint

Our own kit car design for a complete chassis with roll cage. Incorporates the stiffest cage and chassis design possible. Ideal for any sportsman or modified car. Designed by chassis engineer Paul Lamar. Includes all necessary part numbers and material descriptions, plus details of a unique, simple front frame snout. Can be used with conventional coil springs or coil-over-shock units. Plans include two different rear suspensions, two different front snouts, and a brochure with chassis photographs. **Order #B400...$17.45**

Practical Engine Swapping

EVERYTHING you need to know about installing an engine from one vehicle into another. It covers every problem, such as: will it fit • buying hardware • buying an engine • steering and suspension alterations • cooling problems • clutch, throttle and shifter linkage • driveshaft • rear axle swapping • FWD conversions • electrical considerations • engine mounts • and much more. **Order #S111...$5.95**

Sprint Car Technology

A complete book on buying, building and racing a sprinter, midget or super modified. It includes an in-depth study of current chassis design and components, and all the elements required to make a competitive race car. Chapters include: chassis structure • torsion bars – theory – parallel vs. cross – stocked vs. level • straight front axles • live axle rear • De Dion rear • bird cages • lateral locating linkages • steering layout • bump steer • king pin inclination • shocks • roll, roll couple and roll centers • side bite • braking • tires • dirt vs. asphalt set-ups, and more. **Order #S125...$7.95**

Race Car Graphics– The Professional Touch

Takes the "magic" out of race car coloring, painting, numbering, striping, graphic design, etc. It shows you how to make your car and operation look absolutely professional without spending a lot of time and money. Author Gary Smith knows his field – he has designed a number of SCCA and IMSA race cars and two Indy 500 pace cars. 8 pages of full color photos. **Order #S118...$6.45.**

Stalking The Motorsports Sponsor

Want to get a sponsor? This clear, precise info tells you how to reach the hard-nosed businessman and convince him to turn loose with cash. These methods will work for you whether you race on a local, regional or national scale. Tells what has worked for others and which will work for you! **Order #S119...$7.45**

Tech Tips

A monthly question and answer newsletter designed to be your one-stop source for all race car problems, be it chassis, handling or engines. Our panel of experts will handle any inquiry plus pass along new developments. **Order #TT-1... $12 a year.**

Chevy Hot Line

A monthly tech newsletter for Chevy high performance. Primary emphasis is on engines, but there is also hard core stuff on other performance areas. Includes new, hot information from top engine builders, racers and manufacturers all across the nation. **Order #CHL-1...$12 a year.**

The Racer's Complete Reference Guide

This book is the "Yellow Pages" of the high performance world. Tells you who makes it and where to find it for parts, hardware, services, anything. Says **Stock Car Racing Magazine** "The book is a handy guide to the makers and suppliers of parts and services. This book's a bible you can't be without!" New third revised edition just released. **Order #S108... $6.45**

Race Car Fabrication And Preparation

Includes thorough discussions of: prepping a transmission • setting up the rear end • electrical system • building the chassis and roll cage • cutting costs and beating the economy stock rules • welding • clutches • safety systems • wrecking yard parts • fabrication • hardware • wheels and tires • cooling system • plumbing • driveshaft • Plus much, much more. Photo packed, over 160 pages. **Order #S114...$7.95**

How To Go Grand National Racing

A thorough and enlightening book by GN driver Skip Manning. Will help anybody be prepared for a real GN effort. Covers: driving and operating as a rookie • how to make money on incentive, bonus and contingency money • establishing a shop • a complete budget • what you need to buy • where to save money • tricks of the trade • how to prepare a car between races • buying a used and new car and car hauler • and much more. Photo filled. **Order #S122...$7.45**

Racing The Small Block Chevy

An all new book rewritten in Fall, 1979, to bring you completely up to date. Everything you need to get the most power from your engine. Contents: The very latest inside information • tested and proven facts from NASCAR engine builders and Smokey Yunick • revealing info on cams, carburetor mods, head porting • the entire engine & assembly discussed in detail. **Order #S112... $7.45**

Building A Race Car Picture By Picture

A collection of detailed photos taken underneath, inside and under the hood of some of America's most successful race cars. Shows the elements that make a truly winning car: electrical and plumbing detail, component placement and structure detail. Many more topics. Book is all photos (with captions). **Order #S110... $6.45**

Racing Engine Preparation

A.J. Foyt says, "This is the most complete engine book I've ever seen. No nonsense...very thorough." Uses the small block Chevy, Ford and MoPar as examples. Chapters include: understanding camshafts • lubrication system • ignition • complete listing of major part numbers • detailed engine assembly • cylinder head blueprinting • dyno testing. Plus preparation of rods, pistons, carburetors, blocks and much more. 152 pages, 380 photos. **Order #S106... $8.50**